HOW TO MAKE A WEIGHT-DRIVEN CLOCK

HOW TO MAKE A WEIGHT-DRIVEN CLOCK

Richard Simmonds MBHI

THE CROWOOD PRESS

CONTENTS

CHAPTER 1	TIME AND THE CLOCK	6
CHAPTER 2	THE DESIGN OF THE CLOCK MOVEMENT	15
CHAPTER 3	THE TOOLS REQUIRED AND HOW TO USE THEM	20
CHAPTER 4	MAKING THE PLATES AND PILLARS	35
CHAPTER 5	SCREWS	52
CHAPTER 6	MAKING THE WHEELS	59
CHAPTER 7	PINIONS	80
CHAPTER 8	ASSEMBLING THE WHEELS AND PINIONS	87
CHAPTER 9	MAKING THE BARREL ASSEMBLY	95
CHAPTER 10	SETTING THE TRAIN INTO THE PLATES	118
CHAPTER 11	THE MOTION WORK	125
CHAPTER 12	MAKING THE VARIOUS BRIDGES	130
CHAPTER 13	MAKING THE PALLET ARBOR AND CRUTCH	137
CHAPTER 14	MAKING THE PALLETS FOR THE ESCAPEMENT	146
CHAPTER 15	THE PENDULUM	156
CHAPTER 16	FINISHING THE MOVEMENT	163
CHAPTER 17	THE WEIGHT AND PULLEY	168
CHAPTER 18	TELLING THE TIME AND COMPLETING THE MOVEMENT	177
	APPENDIX	188
	INDEX	190

The drawings in this book are not reproduced to scale. The reader should take careful note of the dimensions shown in the drawings' labelling and follow those.

1 Time and the Clock

INTRODUCTION

Clocks and mechanical devices hold a fascination for many of us from when we were young children. In my case I have been fortunate to be able to turn the fascination of small mechanical objects and all things horological into my working life. Since beginning as an apprentice in 1983 I have been privileged to work with many fine clocks in wonderful locations, and have met many interesting people.

Time has shown a well-made clock will give literally centuries of pleasure and good service. Quite a number of traditional weight-driven clocks are well over 250 years old; the oldest I've had the privilege to repair was made in the year 1646. Much history has passed during the life of this particular clock, including most of the industrial revolution!

While much has been written in the past about the traditional methods of clockmaking, here we will discuss producing a clock movement using a mixture of those older working practices (with the required hand tools) alongside more modern techniques and products, where appropriate. Where possible, the use of unique horological equipment has been avoided to bring the project within the scope of a typical workshop setting. Where specialised equipment is required, such as when cutting the teeth onto a clock wheel, alternatives have been suggested to make the completion of the project possible.

As the design and construction of the movement proceeds, there will be opportunity for you to include some of your own personal touches to add interest and uniqueness to the clock movement.

There is a great pleasure in producing a working instrument. Hopefully you will find this clock an interesting project to construct. Perhaps it will be a unique addition to your horological collection, or a starting point into the subject of horology. It might also turn out to be a family clock passed down for many generations to come, and you will have had the pleasure of being the clockmaker!

TIME

Let us begin by thinking about time: you cannot stop it, see it, or smell it – neither can you touch it – but it threatens to take over our lives.

Think about this... what is time? The *Oxford English Dictionary* defines it thus:

> *The indefinite continued progress of existence, events etc., in past, present, and future regarded as a whole. The progress of this viewed as affecting persons or things.*

The author Dava Sobel in her book *The Quest for Longitude* comments about time and our instruments for measuring its passing:

> *Time is to clock as mind to brain. The clock or watch somehow contains the time. And yet time refuses to be bottled up like a genie stuffed in a lamp. Whether it flows as sand or turns on wheels within wheels, time escapes irretrievably while we watch. Even when the bulbs of the hour-glass shatter, when darkness withholds the shadow from the sundial, when the mainspring winds down so far that the clock hands hold still as death, time itself keeps on. The most we can hope a watch [i.e. clock] will do is mark that progress. And since time sets its own tempo, like a heartbeat or an ebb tide, timepieces don't really keep time. They just keep up with it if they're able.*

Early man noticed the sun, and then the phases of the moon, along with the regular cycle of the planets and stars. In horology the phrase 'The Movement of the Heavens' is often used to sum up what we see happening in the sky. The Babylonians began calculating mathematically for the movement of the sun, moon and stars, based on a system known as Sexagesimal, which was 60-based. They loved the number 60, as it could be divided by the greatest number of whole numbers. That is to say, without having any remainders, as they did not know what to do with the remainders. It was the Babylonians who decided there would be 360 degrees in a circle, and 60 divisions in an hour.

THE DAY

The time on earth is set by the movement of the planet around the sun, as can be seen in the image of the 'Orrery'.

An orrery.

There are three types of day that can be measured:

1. The Solar Day. Our lifecycle is regulated by the sun. We can define a day as the interval between two successive transits across the meridian of the sun. The time so determined is 'apparent' or 'true solar' time. This is the time you will see on a correctly oriented sundial. The disadvantage of this time is that the days are not of a constant length. There are two reasons for this. The first is that the orbit of the earth around the sun is not circular but elliptical, so the distance between the earth and the sun changes. The second is that the earth does not move in a plane at right angles to the sun, but in a plane called the ecliptic, which is inclined to the plane of the equator by about 23.5 degrees.

A sundial designed for a hall in Norfolk.

2. The Sidereal Day. For the actual determination of time it is more convenient to use the stars as a reference point for determining the rotation of the earth. For this we use an instrument known as a 'transit instrument'. Greater accuracy is possible when observing a star, which gives a pinpoint of light compared to the sun. There are also many, many stars, so you can make a number of observations through the night. The sidereal day is defined as 'the interval of two successive transits of a star across the meridian'. This day is 23 hours, 56 minutes and 4.09 seconds long, which is consistent in length.

3. The Mean Day. This is the day that we all use. If you imagine a fictitious sun moving about the earth's equator, always keeping the same distance from the earth, you would have a day that was 24 hours long, with all the days being the same length. It is because our mean day has 24 hours that we require the leap year day to correct the difference between the sidereal day and mean day every four years. We work on what is known as the 'civil year' – so 365 or 366 whole days. In addition to the leap year, we also omit three days every 400 years. This is done by omitting three leap years. The year 1600 was a leap year, 1700, 1800, and 1900 were not, but the year 2000 was a leap year. With another adjustment every 400 years, we keep our earth time in step with the heavens.

BRITISH STANDARD TIME

In the UK we use British Standard Time (BST). BST was originally railway time. In the 1840s everyone kept local time, so in Bristol, local noon is approximately 20 minutes later than London, and there is a difference of about 29 minutes for actual local noon between Great Yarmouth and Penzance. In 1847, a National Railway Time System came into being, which was based on Greenwich Mean Time. There was actually a lot of resistance to changing time to a national standard time.

In 1884, it was agreed in Washington DC, USA, at the International Meridian Conference, that the meridian for cartography and charts would be at Greenwich. It was also agreed that the prime meridian would be the same for time and date. Approximately 70 per cent of shipping charts and maps used Greenwich as the prime meridian at this time. Only one nation disagreed with the discussion, finally falling into line some 30 years later.

THE CLOCK

The field of horology is vast, covering anything from the latest quartz wrist watch to the huge clocks found in the great cathedrals and institutions of the world.

The great Astronomical Clock at Strasbourg was first installed in 1352–54 and was an incredibly impressive machine. It was replaced by a second clock designed by the mathematician Christian Herlin, which was completed around 1571 and stands 12 metres tall. Today, we see a third wonderful clock movement, this one by Jean-Baptiste Schwilgué completed in 1843, in the case from the 1571 clock. If you plan to visit the clock, the main automata functions at 12.23pm European Time, which is the local noon in Strasbourg. You need to get there early for a good view.

The earliest records we have of mechanical clocks as we might understand them, come from the monasteries in central and northern Europe in the late thirteenth century. One of the earliest recorded was installed in Dunstable Priory in 1283. These records were mostly in Latin. We understand them to be mechanical clocks, although there is a little uncertainty as the same word – *orlogian*, or *orloge* – is used both for a sundial and a mechanical clock. To be certain it is a mechanical clock being mentioned it is necessary to check through the records for notes about oiling or repairing the orloge.

By modern standards these clocks were very crude; they were made from iron, and very inaccurate, only being able to measure time within fifteen to twenty minutes in a day. It is worth remembering though, since this was probably the only clock either in the monastery or the surrounding area, one would not have been aware of the clock's time keeping. These clocks were weight-driven striking clocks, used to call the monks to prayer. They did not have a dial or hands, so the time would not have been read. Dials didn't really appear until the fourteenth century, and they tended to be divided into 24, rather than 12 segments. When the hand and dial were added the idea was that the hour hand mimicked the sun. If you think about it, it becomes apparent that the term 'clockwise' derives from the fact that the sun travels across the sky from left to right in the northern hemisphere. The division into twelve hours did not happen until another century had passed. These clocks had an escapement and oscillator, which controlled the speed at which they ran. The escapement of a clock is the heart of the clock, if you will. The better the escapement, the better the clock keeps time. The better our hearts are, the better we 'run', so to speak. It is the development of the escapement and oscillator over the years that has increased the accuracy of the clocks we have today.

THE TERMS USED IN CLOCKMAKING

The mechanism of a clock is known as the 'movement' of the clock. The gears of a clock are referred to as wheels, and the smaller steel gears as pinions. When wheels and pinions are set to mesh with each other they form a 'train' of wheels. Each wheel and pinion is mounted on an 'arbor', rather than an axle. On the end of each arbor is a 'pivot'. This is the running surface for the arbor, passing through the plate at the pivot hole, which acts as a plain bearing.

There are a number of different brass alloys available today. In clockmaking, CZ120 is used for the plates and wheels. A clock wheel is generally crossed to save weight. This is done ensuring the wheel is reasonably balanced and pleasing to look at.

Pinions are of a smaller diameter and tooth count. Generally any wheel with less than sixteen teeth is referred to as a pinion. To give better wear, a pinion is usually made from a good quality steel such as EN8 or EN16. This can be hardened and tempered where required.

An example clock movement made by the author.

A wheel with four crossings. This is the more traditional form and is found in the majority of clock movements.

A wheel with six crossings, used in quality clockmaking.

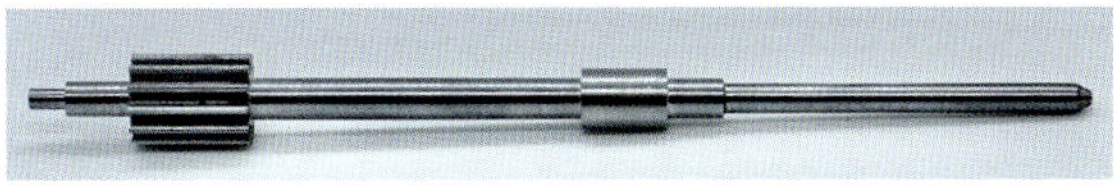

A traditionally turned and manufactured centre arbor. This is used in one of the movements described in this book.

A pinion head before (*right*) and after drilling ready to mount on an arbor (*left*).

Two ways of manufacturing pinions are described in this publication. The traditional way a pinion is formed is by turning the whole arbor complete and then cutting the leaves for the pinion into the arbor.

The second way to produce a pinion is to make the pinion itself as a short section, and then drill the centre of a suitable diameter hole to fit snugly onto an arbor. It is then secured using soft solder or an adhesive such as Loctite 638. The pinions in this project have been designed to have a root diameter large enough to allow this option to be used.

THE FUNDAMENTAL PRINCIPLE OF THE CLOCK ESCAPEMENT

The Escapement

The clock movement is a type of gearbox, and the escapement is the controller or governor of the clock movement. It consists of a number of parts including the escape wheel and pinion, and either a type of pallet or a balance and its associated parts. In this publication a pendulum with its associated parts is used. Finally, there is the crutch, which links the clock escapement to the pendulum.

Whilst the design and quality of the whole movement affects its performance and reliability, careful work when making the escapement will pay dividends with the timekeeping and 'action' as it is referred to, of the escapement. The choice of design for the escapement and the pendulum impact the timekeeping significantly.

The Verge Escapement

The earliest form of recorded escapement is reckoned to be the verge escapement, or in its vertical form the foliot escapement. The origin of this escapement is not known, although it is believed to have existed from the fourteenth century. These clocks tended to have only an hour hand, as their timekeeping was not close enough to require a minute hand. At this point in history there was not the requirement for close timekeeping, with no trains to catch and no *Archers* (a long-running UK radio programme) to listen to on the radio! The verge is a type of escapement known as a recoil escapement.

The Recoil Escapement

In the seventeenth century, the anchor or recoil escapement was developed. Whilst it is not known who actually invented this escapement, Dr Hooke is generally accepted as the inventor, and William Clement the person who actually made it. This is the escapement typically found in a longcase clock, as well as most mass-produced clocks of the twentieth century with a pendulum-controlled escapement. It is known as the recoil escapement, as once the escape wheel has given an impulse to the pendulum, the pendulum continues to swing in the same direction for a short distance, causing the escape wheel to be moved in reverse or 'recoil' before moving forward as the pendulum is driven back in the opposite direction.

An example of the verge escapement. Shown on an English bracket clock, which was not made by the author.

The recoil escape wheel and pallets used in this clock movement. Shown set on a jig to aid construction and final adjustment of the escapement.

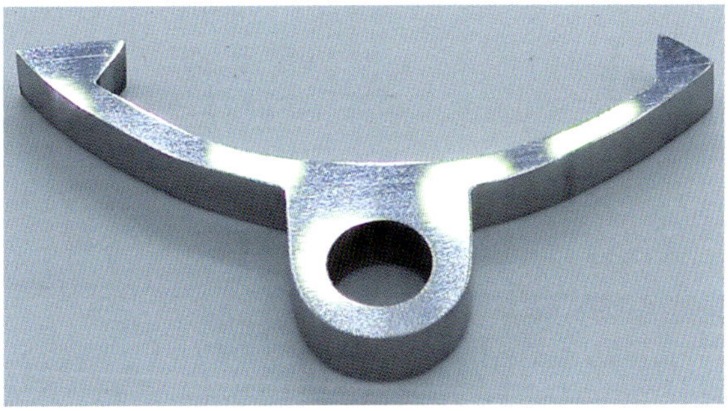

The recoil pallets. Shown finished, but not finally mounted on the pallet arbor.

This escapement gives a satisfactory performance and is not quite as critical in its construction as other more accurate escapements, such as the various forms of the deadbeat escapement.

The Deadbeat Escapement

The deadbeat escapement was developed later in the seventeenth century. Again, it is not clear who invented this escapement, although correspondence between John Flamsteed – the Royal Astronomer – and Richard Towneley suggests that Richard Towneley may have made the deadbeat escapement with Thomas Tompion making a clock using the deadbeat escapement soon after.

When correctly constructed, this escapement is a very close timekeeper. It performs at its best when used with a seconds pendulum – that is to say, a clock that makes one tick or vibration per second, again as a typical longcase clock.

Side note: The correct term for this type of clock is a 'longcase clock'. The terms grandfather, grandmother and granddaughter clocks seem to originate from Edwardian times, when the style and physical size of clock often referred to as a granddaughter clock became popular.

The Grasshopper Escapement

Another escapement that has been shown to work well is the grasshopper escapement designed by the famous clockmaker John Harrison, who was born in Foulby near Wakefield on 24 March 1693 and was responsible for the original development of the chronometer. His grasshopper escapement has been much studied. The Gurney regulator, one of a pair of regulators made by Martin Burgess, once displayed in the city of Norwich, features many of Harrison's innovations. In 2015, the

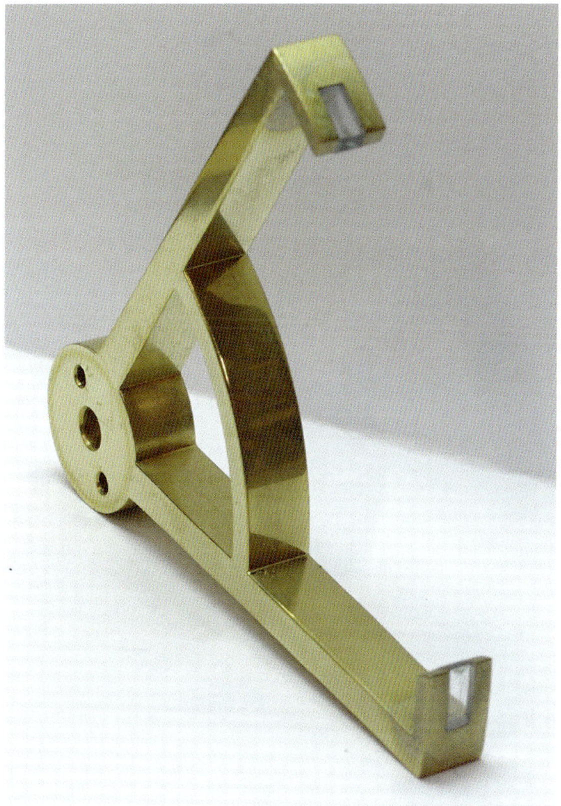

A pair of well-made deadbeat escapement pallets by 'Dent's' of London.

The Brocot style of deadbeat pallet. The assembly consists of three parts: the steel frame and two hardened steel 'pads'. These are inserted into the frame at the correct angle and fixed using Loctite 638.

A variation of the grasshopper escapement constructed by the author. The design shown in the photo above is based on research done in Cambridge, Massachusetts, USA and was made to fit a large regulator where the action of escapement was to be visible.

sister clock to the Gurney regulator took the world record for the best timekeeping for a clock with the pendulum in air, recording a rate better than one second in one hundred days.

The Gravity Escapement

There are a number of other escapements that perform well as weight-driven escapements, such as the gravity escapement, which was developed by Dennison (1840) (later Lord Earnshaw) for the clock in Westminster palace. A variation on this escapement is used by Jens Alson for his fabulous world time clock located in Copenhagen town hall, as it is ideal when the clock has to perform a number of tasks during the course of an hour. It is known as a 'detached escapement', as the wheels driving the clock do not give a direct impulse to the pendulum. This escapement was used in a large chiming longcase clock made by the author and was found to perform well. Owing to requiring an extra wheel in the train, this escapement will not be considered for this project.

HOW THE ESCAPEMENT WORKS

This is best explained by considering one of the escapements we are going to construct: the recoil escapement, shown on the following page. The wheel count and design of the clock described in this publication will be used as an example.

The escape wheel has 30 teeth, and rotates clockwise once in 60 seconds, causing the pendulum to vibrate or beat (tick) once a second – hence the term 'seconds pendulum' is often used. Each tooth of the escape wheel is used twice by the escapement: once as it comes into the escapement – hence the 'entry' pallet – and once as it leaves – the 'exit' pallet. Therefore, 30 teeth will give 60 seconds for one rotation of the wheel. It is a simple matter to add a seconds hand to the front of the escape wheel arbor, and this is why many clocks with a seconds pendulum have an extra dial showing the seconds passing.

The 'recoil' in the escapement comes from the continuation of the pendulum swinging beyond the point where the tooth initially contacts the impulse surface. Owing to the pendulum swinging in a given direction and the shape of the impulse surface, although the tooth has contacted the pallet surface, the inertia in the swinging pendulum means it continues to travel in the same direction a little further, rather than immediately changing direction. The recoil can often be seen when observing the seconds hand on a typical longcase clock. It is wasted energy, an issue the deadbeat escapement is designed to eliminate. Thus, the deadbeat escapement will operate with less force, and has no recoil.

The Action of the Deadbeat Escapement

The deadbeat escapement was designed around the beginning of the eighteenth century. It seems likely that Richard Towneley was the first to make a form of this escapement. As an escapement it requires very careful construction. For this project the version of deadbeat escapement constructed is known as the Brocot pin-pallet deadbeat escapement. The author has chosen this version as it is more readily constructed. Whilst the traditional design is superior, the Brocot escapement performs very well as a seconds escapement.

THE ACTION OF THE RECOIL ESCAPEMENT

1

Tooth A comes into the escapement and is resting on the entry impulse surface of the pallets. As the pendulum swings, this tooth slides along the pallet impulse surface until it drops off the end of the impulse surface.

2

This releases the wheel, allowing it to turn. Note that the small brass tube (bottom right) shows the location of the pendulum.

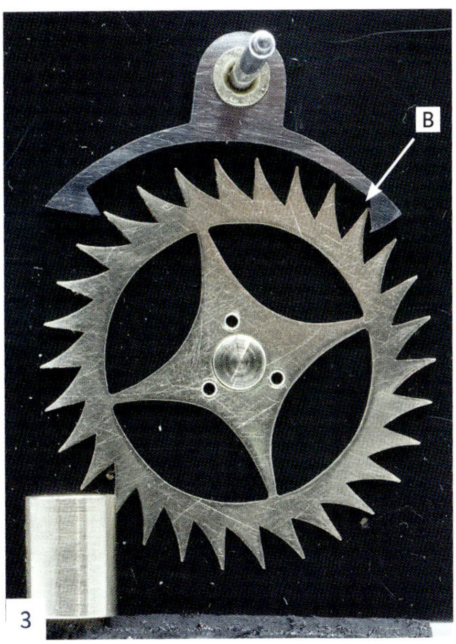

3

The rotation of the escape wheel is brought to a stop by tooth B coming into contact with the exit impulse pallet. On contact it imparts an impulse, driving the pendulum back.

4

In turn tooth B moves along the impulse plane until it runs off the end of the exit pallet, allowing the wheel to turn. The rotation is stopped by tooth C coming into contact with the entry pallet and thus the action is repeated.

THE ACTION OF THE BROCOT PIN-PALLET DEADBEAT ESCAPEMENT

The left hand impulse pin – the entry pallet – in contact with the tip of the escape wheel tooth A. As the pallet moves to the left, tooth A will follow round the curve of the pin imparting impulse to the pendulum, and tooth B will head towards the exit pallet.

Tooth A now falls off the entry pallet and the escape wheel rotates until tooth B locks onto the exit pallet, as can be seen in the photo above. The pallets will continue to move to the left a little longer but the wheel does not recoil. The pallets then begin moving back to the right with the exit pallet giving impulse to the pendulum. Tooth B is released by the exit pallet and the escape wheel turns until tooth C comes into contact with the entry pallet and the action begins again.

The Design of the Clock Movement

2

WHILST THE FUNDAMENTAL DIMENSIONS and construction must be followed for the clock movement to work, do take the opportunity to shape some of the parts to your own design when the text indicates. This will give you the satisfaction of owning a unique clock in which you have been part of the design process.

There are two options for the movement to be wound – either using a winding key, or by pulling a brass chain to lift the driving weight to wind the movement up. There are also two designs of escapement described, the recoil and deadbeat. It is a good idea to settle on which variation of movement you are going to make beforehand, as there are differences as the movements progress.

WINDING THE MOVEMENT

Weight-driven clocks require winding by lifting a weight, which, as it falls owing to gravity, drives the clock. Traditionally clocks of a shorter duration, usually of 30 hours, tended to be wound by pulling a chain or rope to lift the weight. During the nineteenth century, chain-wound quarter-chiming longcase clocks of eight days' duration were produced. Many mass-produced longcase clocks produced today are also chain wound. The chain-wound barrel is the more straightforward of the two to construct, and does not require a hole through the dial for winding.

Please note that to illustrate this publication, the same set of clock plates is used to show the chain- and key-wound barrels. The chain-wound barrel is shown on the left of the movement, but can be located on either side.

The alternative method of winding a weight-driven movement is to use a key.

A key-wound barrel has a more refined feel in use and as a design is *generally* associated with clocks of better quality and longer duration. It allows durations of up to a year to be achieved within the bounds of a typical longcase clock case. It is most often used in clocks of eight-day duration and this is the duration used for this project. Whilst its construction is more involved, it is very satisfying to use.

The chain-wound movement. The chain barrel is mounted on the left-hand side.

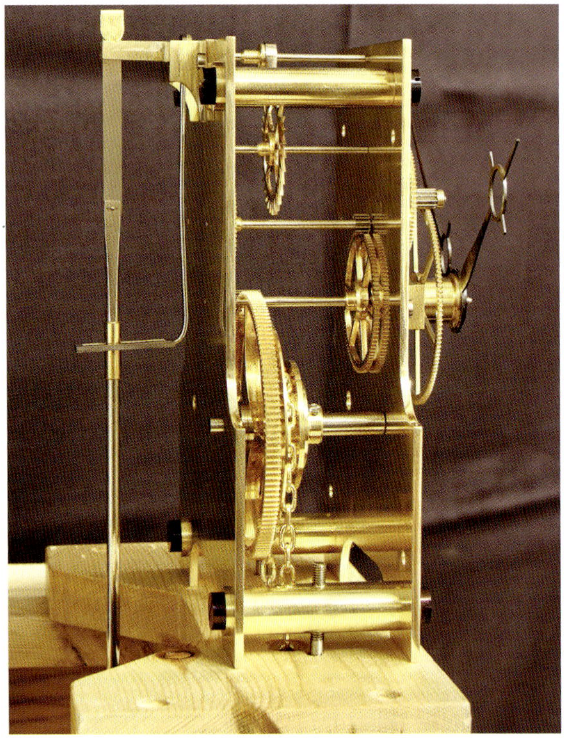

The chain-wound movement seen from the left-hand side showing detail of the chain on the sprocket.

The key-wound barrel with the barrel mounted on the right-hand side. Both designs of barrel can be mounted either side of the movement, depending on the design of the dial.

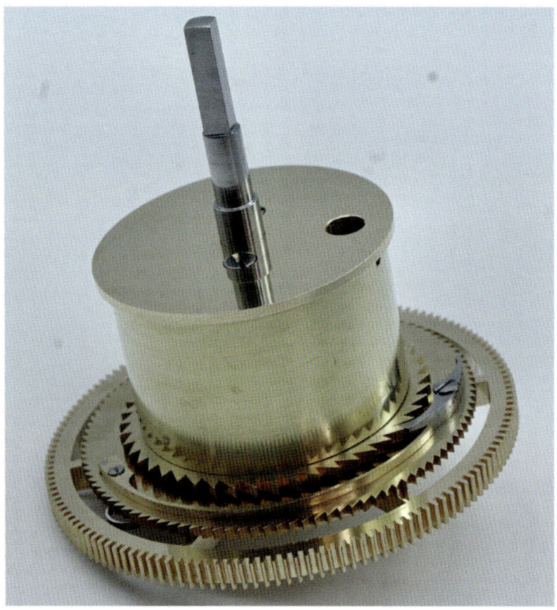

The complete chain-wound barrel assembly. It is more straightforward to construct than the key-wound barrel.

The complete key-wound barrel assembly.

The key-wound barrel assembly viewed from the front of the movement.

THE ESCAPEMENT OPTIONS

The first option is to make the clock with a recoil escapement. This is perhaps the best place to start if you are new to horological construction projects. This escapement performs well, and is reasonably straightforward to construct. Note the same train of wheels can be used to make the movement with either the recoil or deadbeat escapements.

The second is to make the movement with the deadbeat escapement. This requires a little more care in construction. It does give better timekeeping and is perhaps a little more interesting to construct if this is not your first horological project. If the traditional deadbeat escapement were used, there would be the need to introduce a device known as a 'maintaining power device' – unless you *without fail* were to stop the clock before winding. The maintaining power device is an interesting addition to the construction of the movement and is shown overleaf.

Notes and drawings are included for construction of the movement with the Brocot deadbeat escapement.

As the Brocot deadbeat design is used here, the maintaining power device need not be constructed, but the movement would be better for its inclusion. The construction of the maintaining power device will be covered in due course. In addition to space for the weight, there will need to be room for the seconds pendulum to swing beneath the movement.

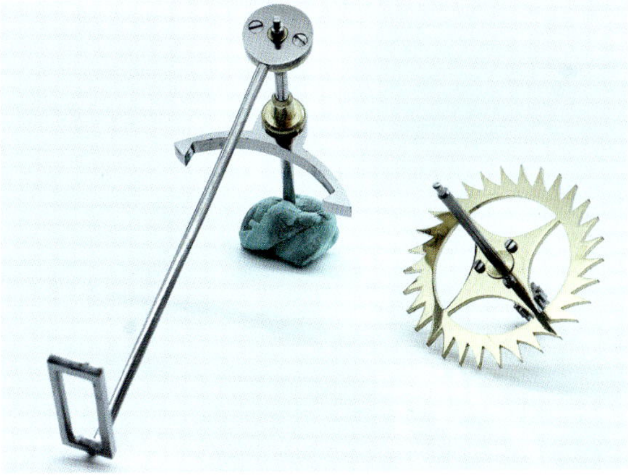

The finished parts of the recall escapement. The escape wheel is crossed in the traditional four crossing pattern.

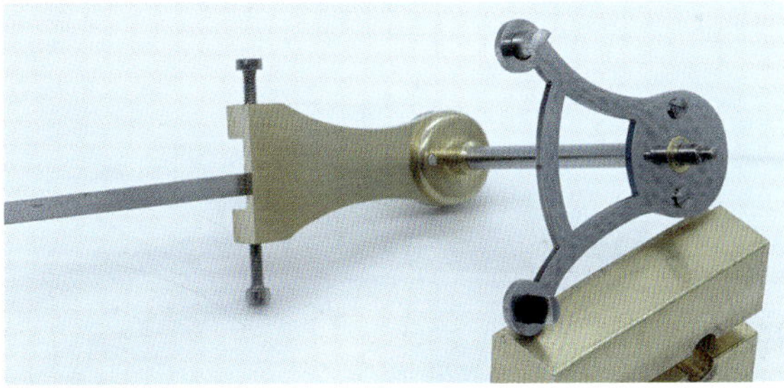

The deadbeat pallet assembly shown with the beat adjusting crutch.

As the clock is driven by a weight rather than a spring, there will need to be room under the movement for the weight to fall. An alternative would be to use a pulley system to route the weight above the clock. This would be an interesting feature, and is how turret clock weights are sometimes located in order to get the maximum height for the weights to fall, and therefore the longest period of time between winding of the clock. The type of clock found in church towers and town hall towers and the like, are referred to as turret clocks.

Wheels and pinions with higher counts (that is, with more teeth) work more smoothly than those with low counts. For example, a pinion with seven or eight leaves (teeth, but referred to as leaves when describing a pinion) will work perfectly well and the majority of traditional longcase and bracket clocks have these, but a pinion of twelve leaves will work much more smoothly. Clocks known as regulators have higher train counts. On the following page is a comparison between a typical longcase clock and a smoother running equivalent train. The list also includes the correct term for each wheel.

The key-wound barrel with the extra maintaining power ratchet wheel fitted.

The reverse side showing the steel spring used to drive the movement whilst the clock is wound.

WHEEL TERMS AND DIMENSIONS

WHEEL/PINION	TYPICAL TRADITIONAL CLOCK TRAIN	HIGHER COUNT TRAIN: SMOOTHER RUNNING	THICKNESS OF WHEEL	DIAMETER OF WHEEL/ PINION BLANK FOR HIGHER TRAIN COUNT	MODULE OF TOOTH (WHICH CUTTER IS REQUIRED)	CENTRE HOLE DIAMETER FOR MAKING THE MOVEMENT
Great wheel:	96 teeth	144 teeth	5mm	88.05mm	M = 0.6	8.0mm
Centre wheel pinion:	8 leaves	12 leaves	n/a	8.17mm	M = 0.6	N/a
Centre wheel:	64 teeth	96 teeth	2mm	49.38mm	M = 0.5	8.0mm
Third wheel pinion:	8 leaves	12 leaves	n/a	6.8mm	M = 0.5	N/a
Third wheel:	60 teeth	90 teeth	1.5mm	46.38mm	M = 0.5	6mm
Escape wheel pinion:	8 teeth	12 leaves	n/a	6.8mm	M = 0.5	N/a
Deadbeat escape wheel:	30 teeth	30 teeth	1.5mm	38mm	Deadbeat cutter B	6mm
Recoil escape wheel:	30 teeth	30 teeth	1.5mm	44mm	7E recoil	6mm
Cannon pinion:	48 teeth	78 teeth	2mm	40.4mm	M = 0.5	6mm
Reverse minute wheel:	48 teeth	78 teeth	2mm	40.4mm	M = 0.5	6mm
Reverse minute wheel pinion:	6 leaves	12 leaves	n/a	6.8mm	M = 0.5	n/a
Hour wheel:	72 teeth	144 teeth	1.5mm	73.38mm	M = 0.5	8mm
Ratchet wheel:	48 teeth	48 teeth	6mm	58mm	Straight ratchet cutter	8mm

EXTRA READING FOR WHEEL-CUTTING THEORY

As we continue to discuss the design of the clock, I would recommend J.Malcom Wildes' excellent book *Wheel and Pinion Cutting in Horology* (The Crowood Press) to give you a theoretical background into the whys and wherefores of clock train design.

The aim is to have an efficient and smoothly running train. The dimensions given in the column marked 'Higher train count' are the ones used to make the movement in this publication. The figures given in the typical clock train come from many individual wheel counts taken by the author over a number of years.

Note: when purchasing materials it is sometimes easier and more cost-effective in the UK to purchase the imperial thickness – especially brass sheet. For example, 3mm CZ120 brass may be more costly than 3.175mm (also known as 10 gauge or 0.125"). If you choose to do this, remember to add or subtract the difference shown in any relevant dimensions.

BRASS

There are several types of brass available for purchase, all of which have specific uses. Brass is characterised by its alloy properties, which are altered by changing the ratios of copper and zinc and by adding other elements such as tin and aluminium. Changing the makeup of the alloy alters how readily the brass is machined as well as how easily it bends and corrodes. The colour also varies with the makeup of the alloy.

For clockwork, CZ120 – also known as half-hard or engraving brass – is used for plates and wheels. It is best used for all parts of the movement, although CZ131 bar can be used for the plate pillars and wheel collets. This is slightly more yellow in colour, and requires more care when drilling. In all notes for working and finishing all the brass parts for the movement it is assumed the maker has used CZ120. (CZ131 finishes in the same way). Do not be tempted to use another grade, as they all have different machining properties, and can be more difficult to work with. For sources of brass and other materials, *see* the Appendix. In the interest of research for materials for this publication the author has purchased brass from a number of sources detailed as being CZ120. From experience, UK-sourced brass made within Europe and the UK is by far the best for clockmaking and is readily available from the suppliers listed in the Appendix.

We now have the basic dimensions for the clock movement. Let's now consider the tools required and how to use them.

The Size of the Clock Plates

Two brass clock plates of CZ120 Brass 3-4mm thick and 150mm by 200mm are required. This size of brass plate blank will allow either design of plate shown to be produced.

The Pillars to Separate the Plates

Four pillars are required, two of diameter 12mm and 59mm long, and two of diameter 15mm and 59mm long; in both cases, these can be machined from a single length of bar stock length 150mm for each diameter.

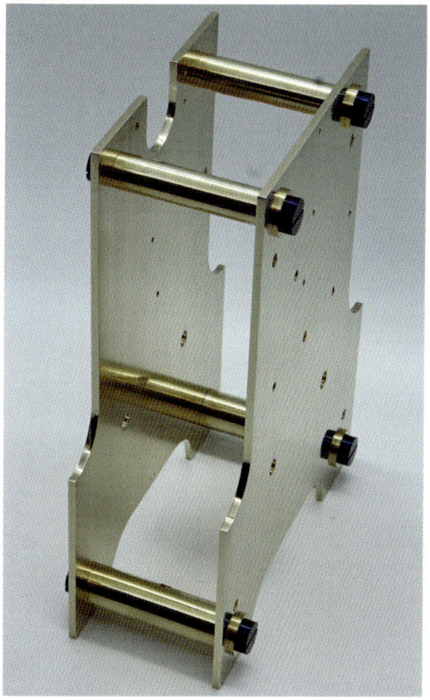

The clock plates finished and secured the correct distance apart by the four pillars. The bottom two pillars are of a larger diameter to allow securing bolts to be screwed into the underside.

3 The Tools Required and How to Use Them

THERE ARE A NUMBER of hand tools you will need to make the clock. You will also need a certain amount of machinery – a lathe and pillar drill are required. Try to avoid modern low-cost tools. These tend to be poorly made and make it difficult to achieve good quality work. Buy once and buy well: good second-hand tools are often a well-priced option.

Whilst a workshop equipped with several machine tools is the ideal, this project can be made with just a pillar drill and lathe. Please note that safety guards should be fitted to machinery and used. They are often not shown in the following photographs as they would obscure much of the details required – but please fit and use them along with safety glasses and other protective equipment suggested in machinery handbooks.

Some furniture is also required – a solid workbench at a height you can comfortably sit at with a good posture and a suitable chair. An MDF or plywood working pad that you can put on the workbench in front of you to work on, cutting, drilling and so on, without damaging the bench top, is also a good idea. Approximately 300 × 500mm is a good size. Check there are no edges to pick up a splinter from.

You will also require good lighting above your bench and/or a large window. The key to seeing clearly is good light.

There are plenty of views of the author's workbench as the book proceeds.

THE TOOLS

Pillar Drill (sometimes referred to as a drill press)

A good solid drill is required. The author's elderly but solid 'Fobco' is illustrated. A good range of speed is helpful, as a slow speed is best for drilling brass.

Fobco 'Star' pillar drill.

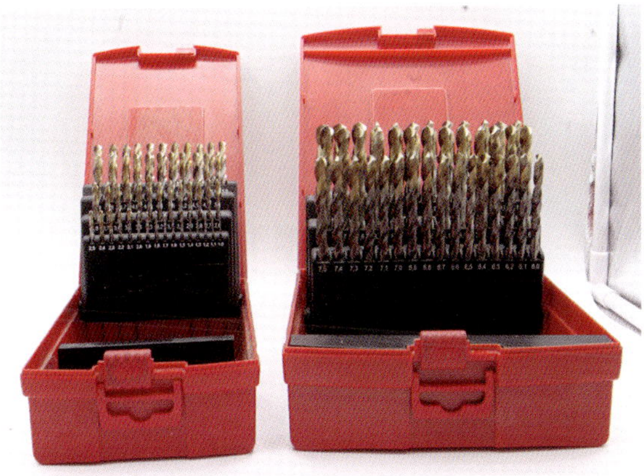

Two sets of Dormer twist drills. The one on the left 1.6mm in 0.1mm steps. The right hand set extends the range from 6 to 10mm in 0.1mm steps.

The different shape of a centre drill is clearly seen.

Drills

A box set of metric 'engineer's jobbing' drills from 1 to 6mm is required, ideally in 0.1mm steps, suitable for drilling brass. A set of regular jobbing drills can readily be altered using an oil stone to ensure they will cut brass smoothly. Adjusting the drill to cut brass smoothly is explained in Chapter 4, in the section 'Drilling Brass'. Once altered, use these drills just for brass, so have an un-altered set too.

Centre Drill

A selection of centre drills is required. Just two or three are ample for this project. A centre drill has a different-shaped tip to a regular drill, and was originally designed to drill into the end of a bar stock in the lathe or mill to allow a tailstock centre to be fitted. It is also useful to ensure a twist drill locates centrally when drilling a hole in the lathe. Spotting drills are another type of drill that work very well when centralising a drill onto a centre-punched hole in the brass movement plates, etc. Both types are shorter than a typical drill, are more rigid, and tend to wander less. The author has used centre drills in this publication.

Oil Stone

This is a flat stone, usually two-sided, having a coarse and smoother side. It is used for sharpening and altering cutting tools. It can also be used for finishing a surface to be flat. When using it lubricate it with a thin oil, hence its name – oil stone.

A selection of tools listed. The oil stone for sharpening tools is shown on the right-hand side, with a 'graver' turning tool positioned on top.

Lathe

A good-quality, second-hand lathe is better than a 'bargain' modern low-cost machine. The author's lathes are illustrated. As can be seen, they are definitely not new, but well loved and maintained pre-owned machines, still running very well indeed and capable of good work. There are many new lathes of satisfactory quality, of course. With the lathe you will need a certain amount of tooling. In this publication, collets are used for holding the workpiece in the lathe. A self-centring three-jaw chuck is also

Schaublin 102 lathe.

Schaublin vm102 lathe. This lathe is equipped with digital readout scales speeding up the process of turning.

Pultra 1770 lathe and three-jaw chuck. This lathe also accepts collets.

Three- and six-jaw chucks. A chuck is a tool with jaws that close around the object to be turned. These ones are self-centring – as they are tightened the jaws close at the same time, meeting in the middle.

A 'collet' is a collar divided into three equal sections, which is tightened around a section of bar stock the same diameter as the hole in the collet. It is a very accurate way to hold a round bar for turning. Two are shown: the smaller is a 10mm thread used in the Pultra1770, and the larger a W25 Schaublin collet. W25 indicates the diameter of the collet thread being 25mm.

The four types of cutter used for this project. Starting top left and then clockwise: recoil escape wheel, ratchet wheel, deadbeat escape wheel, and bottom left a clock train wheel.

used. A luxury item for the lathe is a self-centring six-jaw chuck. Three different shapes of cutting tool are used: a lozenge shape for general turning; a round-nosed for decorative turning; and a parting tool for cutting – parting – the workpiece off the bar stock in the lathe. The use and shape of these is discussed as they are used.

All these lathes are used with collets and chucks. There are many alternatives to these brands, of course: Myford and Sherline, for example.

Horological Wheel Cutters

These are from P.P. Thornton Successors. The photo shows the different forms used for the different types of tooth profile used in a clock movement. Chapter 6 deals with the various types and how to use them.

The cutters required for this project are:

The ratchet for winding the clock requires a ratchet wheel cutter, and there is no module number.

For cutting the great wheel Module 0.6.

For cutting the centre, third and hour wheels Module 0.5.

Depending on which escapement you are making you require either a number 7E recoil or a deadbeat cutter 'B'.

The centre arbor pinion requires a pinion cutter of 12 teeth and Module 0.6.

The 3rd, escape and reverse minute wheel arbor pinions require a pinion cutter of 12 teeth and Module 0.5.

The Piercing Saw

A piercing saw is a very useful and straightforward tool to use. As with all straightforward tools, its usefulness will depend on mastering the technique to use it. There are quite a number of blade cuts to choose from – very fine to rather coarse. For this project, two grades of blade will suffice, known as 02 and 2. Do not be tempted to buy cheap blades, as they will not last and you will become frustrated very quickly! 'Gold-shark' is a good brand when it is

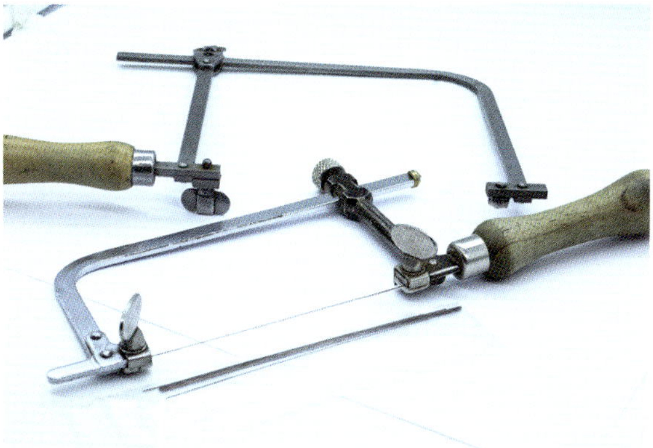

The two sizes of piercing saw used for the project. The plastic pack contains ten replacement blades.

available. Two sizes of piercing saw frame are readily available; purchase one of 65mm and one of 150mm depth. Set the jaws to be approximately 114mm apart.

Saw blades are set in the frame with the teeth pointing towards the handle as it cuts on the down stroke; the action of the saw is up and down – not forwards and backwards. The cutting stroke is the down stroke. The saw is used by placing the workpiece – the section you are working on – flat on the work mat, on the bench and pulling the saw down; thus, the cutting action helps to hold the workpiece down on the bench. It is also easier to see exactly where you are cutting – you can see the blade actually cutting through. Each careful stroke counts towards a closely cut finished piece.

SAWING EXERCISE

Take a piece of 1.5mm thick brass sheet, and using a photocopy of the pattern provided for practice, cut between the lines as closely as you can. The parts should fit together closely when you have cut them out. Persevere before moving on: taking two or three complete attempts will save you frustration when working on the clock itself. Blades will be snapped, cheap blades much more so than quality items.

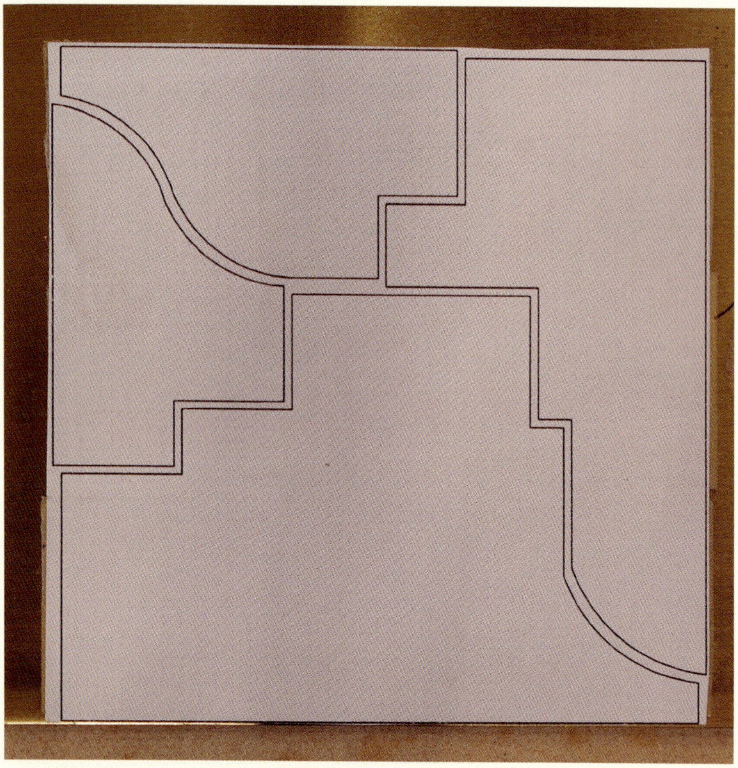

Copy this pattern and secure it to the brass using double-sided tape.

The saw mid-cut.

The neater the cut, the less filing to finish is required. Filing is also a skill to be learnt, and so the neater the cut the less skill (patience!) you will need with the file later.

Insert a saw blade – let us say, grade 02 – between the clamps, making sure the teeth point towards the handle of the saw. For a standard-length piercing saw blade, set the frame jaws 114mm apart (measuring the inside of the jaw) and secure one end of the blade. Next, apply a small pressure to the frame, slightly reducing the gap between the jaws of the saw and secure the remaining saw blade end so that it is under tension. A little practice will soon guide you to the correct tension for your saw frame. (There is not a standard format to measure this.)

Hand Files Large and Small

Small files are referred to as needle files. A set of six needle files with the following files would work well for making the clock's movement: round; half-round; pillar; flat; square; and triangular.

Larger files tend to be referred to as hand files; sometimes the coarseness of cut is included in the name. For example, this is a flat number 2 cut file.

Purchase two hand files of number 2 cut: one flat with a safe edge, and one half-round file. Also purchase the same shape files but with number 4 cut; these are finer and will leave a smoother finish.

Note that for Swiss files, the fineness of the cut is indicated by a number. The higher the number the finer the cut. So, a file numbered 4 cuts with a finer cut than one numbered 2.

Filing is another skill well worth practising before beginning the clock. Using the sections cut out in the piercing saw exercise use the flat file to 'file' all the edges flat, removing the saw cuts. Note that one edge of your file should have an edge with no file teeth; this is known as the safe edge, that is one without any cutting teeth formed on it. Take care with the safe edge of the file; use it towards the area not needing filing, so you do not mark or remove any more material than required.

Hold the file with your dominant hand on the handle and use your other for steadying the leading end of the file. A file cuts as you push it away from you; take light strokes, and do not press too hard. Do just two or three cutting strokes and check to see if you are filing level – that is to say, straight across the workpiece – and following your line. The photo below illustrates the action of filing.

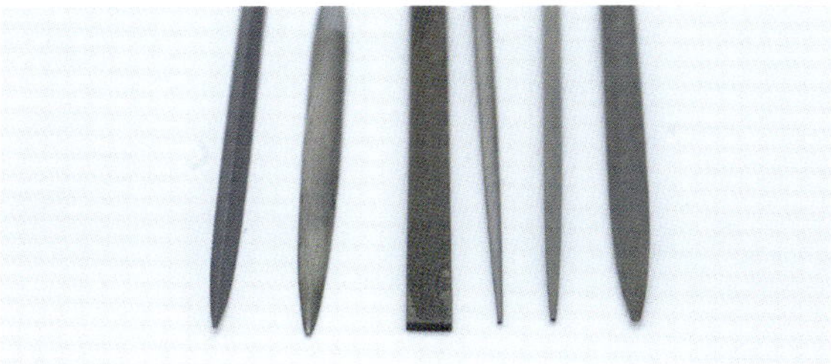

Selection of needle files.

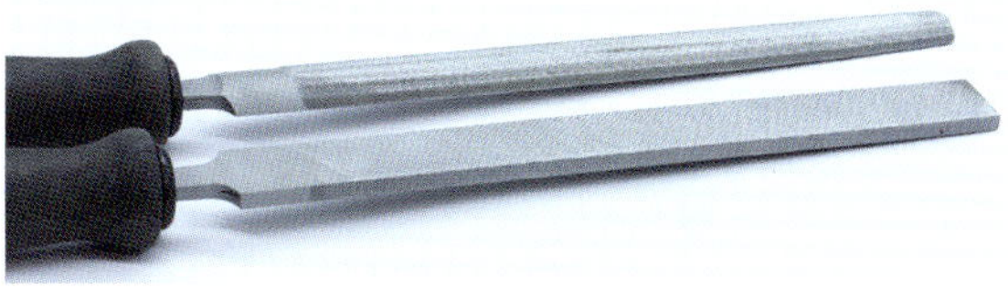

Two standard hand files.

Holding the file with your dominant hand on the handle.

Working a piece of thin brass through the teeth of the file, pushing the clogging out of the teeth.

It can be seen from the photo that the brass deposits in the file have been removed, but there's a bit more cleaning out still to do.

Shown in the photo is holding the flat No. 4 cut file to draw-file the edge where the surface has previously been filed across the surface.

Check often to see that you are following the pattern lines. It does take practice to file leaving a flat rather than a rolled edge.

Next, use the round side of the half-round file to tidy the curved edges.

From time to time the file teeth will clog. Do not use a 'file brush' to clean out the teeth, as these take the sharpness from the teeth of the file. Instead, use a thin piece of brass (anything in the way of an off-cut is fine), and push it along the line of teeth on the file. This will push the clogged metal from the teeth. Leaving the teeth clogged will ruin the file and spoil your work. You may also find it useful to lubricate the file with white spirit while you clean it out. White sprit is also known as mineral spirit, mineral turpentine, and stoddard solvent.

The file can also be lubricated with white spirit, or charged with French chalk to help stop it clogging whilst you use it, although this makes the operation rather more messy. Generally when filing newly cut brass, the author tends to lubricate the file with white spirit, as it largely stops the clogging of the teeth.

Some aspects of the sawing exercise are better filed to shape using a needle file from the selection purchased. Placing the handle of the file being used into a pin vice to act as a larger handle will make using it more comfortable.

The files so far have been used by pushing away from you to cut and pulling back ready for the next cut. Another technique when using the file is to 'draw file'. As can be seen in the photo, this is done by placing the file on the workpiece, and holding either end of the file, drawing it along the edge previously filed. This gives a neater finish following the initial filing process. Standing up can often help when filing generally, more so when draw filing.

The Graver

This is a tool that is generally used when using a small lathe for turning. It is a hand-held turning tool, but has another very useful function when finishing the edge of your work. There are several shapes of graver available; purchase a lozenge-shaped item.

A final finish to practise here is to tidy the draw-filed edges of the exercise by using the edge of a lozenge-shaped graver – a hand-turning tool – held in a pin vice and drawn along the edge of the work gently scraping the surface, without applying too much pressure. With practice, this will give a smoother slightly polished edge.

Using the graver to scrape the edge, giving a smooth finish.

The author gave this exercise to a trainee horologist to see how it worked. As a guide for what you are aiming for, the photos on page 27 show a satisfactory result.

On the left is the complete exercise, and on the right a close up of the fit of pieces.

Screwdrivers

Screwdrivers are straightforward to use. It is important to use the correct size blade for the screw being adjusted. In horology, straight or slotted screw heads are used. There is no fundamental reason why cross head or Allen head screws should not be used in this project, except that the appearance of a slot-headed screw is more elegant, and can be shaped to suit your own personal taste quite easily. We are aiming to make a clock movement rather than a piece of engineering! As the screws tend to be steel, and the object they are being screwed into tends to be brass, it is possible to tighten them sufficiently with a slotted head. In addition, screws of dissimilar metals are less likely to bind in the threads.

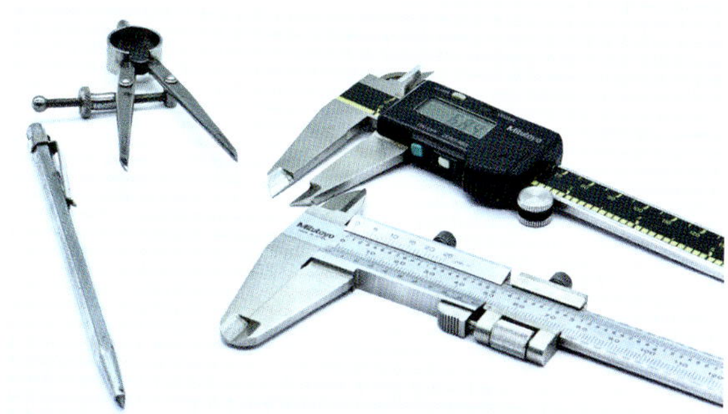

Scribe and engineer's dividers left and vernier gauge – digital and analogue versions, shown right.

Measuring and Marking Tools

The vernier gauge, steel rule, dividers, engineer's set square and scribe are used for marking out and measuring when making a particular part. A micrometer is not essential for this project. Purchase a steel rule of 300mm length, with the first 100mm divided into 0.5mm graduations. A shorter rule of 150mm is also useful, as it is less clumsy to use for small measurements. All measurements given in this publication are in millimetres.

One further marking-out tool is useful when marking metal. This a large permanent marker pen or a tin of blue colour such as Dykem Steel Blue Layout Fluid. In either case, the colour is used to create a dark uniform and thin surface so that when a marking line is scribed it shows as a bright line, making it much easier to follow. More on this when we come to use it in due course. In this publication, a large blue permanent marker is used as they are clean, quick to use and easily available.

Vernier callipers are marked in inches and millimetres. Generally, the upper scale is imperial and the lower metric. We will concern ourselves with the lower scale, which is metric and 0.1mm in fact. To read the vernier gauge open the calliper so that it fits easily over the object you wish to measure.

To illustrate, the photo below shows measuring across the diameter of a piece of round bar stock. Close the calliper round the bar until both sides are in contact with the bar. Looking at the scale marked 0–10 on the lower side of the sliding calliper, the 'zero' gives you the whole number, so looking at the photograph it is 12. The measurement will therefore be 12 'point something', so to speak. Each graduation on the lower scale is a line mark numbered 0 to 9. Looking closely at these lines, and the graduations on the main non-sliding scale, you will notice the lines begin to line up with the sliding calliper lines. Whichever line is the closest match will be the decimal reading: in the photograph it is 6. So the final reading to 0.1mm is therefore 12.6mm.

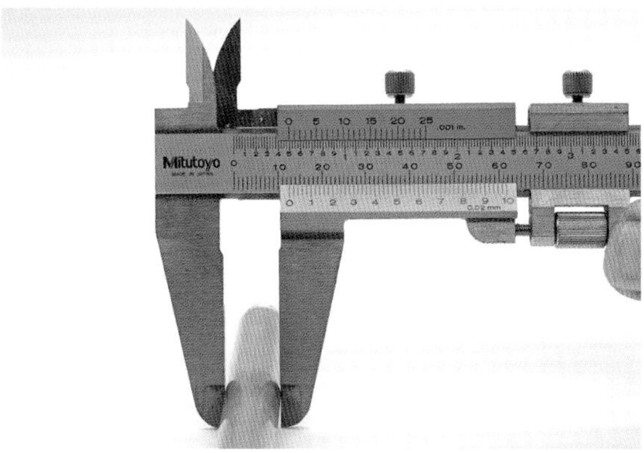

The analogue vernier showing 12.6mm as the diameter of the brass bar.

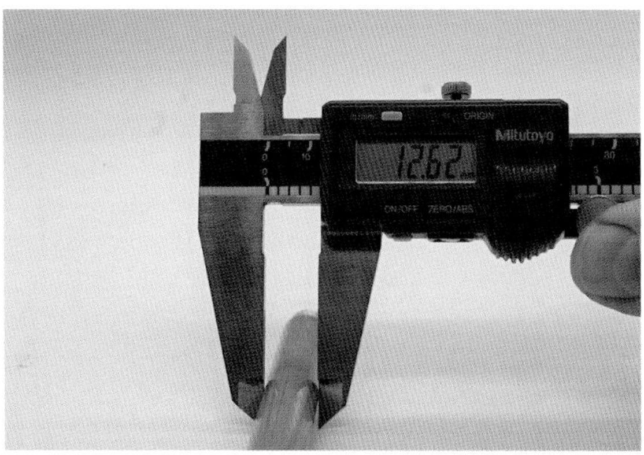

The digital vernier confirming the measurement.

There are a number of digital vernier gauges available; please only purchase a good-quality item. Shown below are the author's Mitutoya analogue and digital vernier gauges.

These are the tools used for measuring and marking. When marking out on a sheet of brass note that when purchased, it ought to have at least one true or straight edge. Use this edge to measure and mark from, then the set square can be used to ensure parallel lines for the movement plates, and so on.

The Hammer

The small hammer as illustrated is quite sufficient for most tasks required when making the clock. It is surprising how wrong something can go even with a hammer of this size and a large hammer will not help at all! During the course of making the clock it will be necessary to work-harden brass to make a spring. For this a larger hammer can be used. The hammer is double-ended: the flat end for hitting a punch, hardening brass for making springs, rivets and so on, and the wedge-shaped end for riveting objects together.

Bench Anvil

An anvil of a size that will easily stand on the workbench is useful when riveting two parts together or for work hardening brass to form a spring.

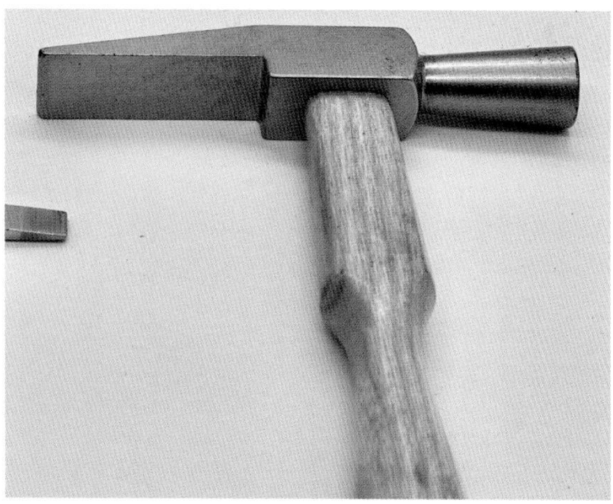

The double-ended clockmaker's hammer.

Bench-top anvil.

Selection of Four Pin Vices

Eclipse No 121, 122, 123 and 124, and one hand vice, which is similar to a pin vice, but has the jaws of a vice. Shown is a set purchased second-hand online. These are considerably better made when compared to typical modern items and feel much better to use. Pin vices are for holding small diameter circular objects to work them.

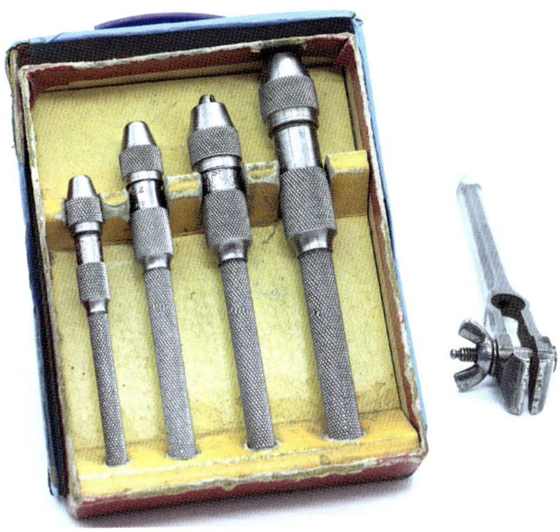

Selection of four pin vices and one hand vice.

Among the tasks you will use one for is pivot polishing, and filing taper pins. They are also useful as needle file and five-sided broach handles. The hand vice is like a pin vice in operation, but is used to hold objects with flat sides.

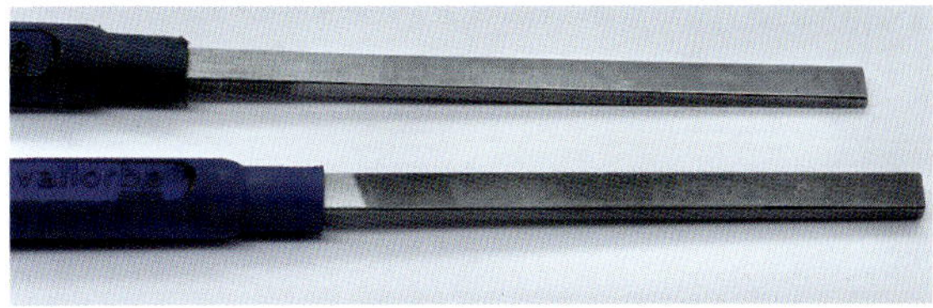

The burnisher/file is a double-ended tool, both ends of which are shown in the photo.

Pivot File/Burnisher

This is a tool traditionally used in clockmaking for finishing pivots. They are expensive and easily spoilt, so use it only for its specific task, and lubricate it with white spirit. As with the other hand tools listed here, whilst it is a quite simple tool, its usefulness comes through mastering its use. For an exercise to learn how to use it, try practising the following. Take the regular fine cut – Swiss 4 – flat file, and keeping the safe edge towards the pin vice, try filing some steel pins. These pins will file much more quickly with a number 2 file, but use the finer 4, as this will have little more of the feel of the very fine pivot file.

Take a piece of very hard wood approximately 25mm wide, 80mm long, and 30mm deep (ebony, lignum-vitae, or a hard mahogany is best). Cut one end at an angle, not too steep, and then place the wooden block in the vice. Into the top surface use a sharp workshop knife and cut a small groove along it. Take a piece of regular mild steel approximately 2mm in diameter – a piece of old wire coat hanger, circular nail, or similarly-dimensioned mild steel wire will work well – 40mm or so long and mount it in the pin vice of nearest size. Now take a regular flat file and, taking care to place the safe edge towards the pin vice, file the diameter of the steel down until you get to approximately 1.5mm diameter. Do this by placing the file lightly on the steel and turn the pin vice towards you whilst pushing the file away from you. (*See* photos on the following page.)

The photo on the top left shows the file placed on the pin ready to begin, and the one below it pushing

The starting position for filing a pin.

the file away whilst rotating the pin vice and hence the steel pin towards you.

At the end of the stroke (top right), turn the pin vice away from you whilst pulling the file back towards you. It will take some practice to obtain a good smooth action. Persist until you have a smooth action and can produce a 1.5mm-diameter pin that is *reasonably* round. It will not matter to begin with if the pins have a slight taper going slightly thinner than 1.5mm at one end, as this is the shape of a traditional tapered clock pin. We will use a few tapered pins later in the project. However, for finishing the pivots in due course these will need to be true.

Practice is the name of the game: do at least ten of them in a practice session, and do several practice sessions until the pins are round, smooth, non-tapered and a consistent size. Place twelve of the best of these pins to one side as we will use them later.

This is the same action you will use to finish the pivots for the clock. It is better to master the technique with an exercise than to spoil the pivots of a nearly completed arbor. In addition the pivot burnisher/file is an expensive file to purchase, so learning the tech-

The end of the stroke.

nique to use it before actually using it and without damaging it is worthwhile. Remember the pivot file and burnisher are lubricated, so try lubricating the file for this exercise.

The Five-Sided Broach

Clockmakers' broaches are used for finally fitting the pivots to the bearing hole. A clock is a precision instrument, but unlike a piece of general engineering it is not made to be a 'tight' fit. All the parts have to be well made in order to fit and work together as a piece of clockwork. A five-sided broach helps to fine-tune the fit of the pivot into the pivot hole. This process is also a smooth cutting process, and so the pivot hole is left relatively smooth inside. Broaches can be purchased

This is a large five-sided broach to show clearly the shape.

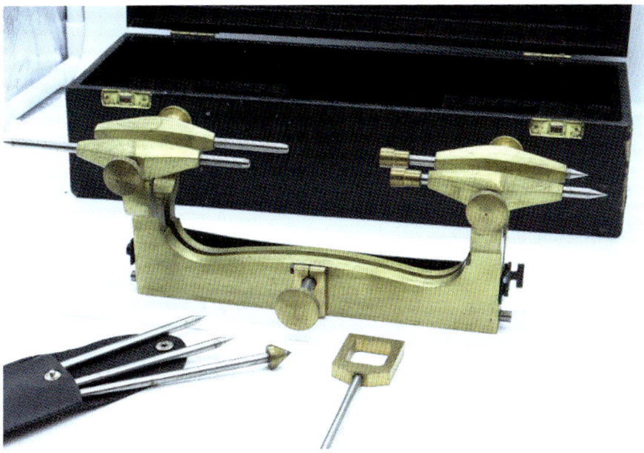

A traditional clockmaker's depthing tool. There are other variations on the design, one of which is used in due course.

as a set of twelve, and it is suggested you do this as a cost-effective way of gaining a good range. The use of the broach will be described when pivots have been turned and are ready to be fitted in the pivot holes drilled in the plates.

Depthing Tool

Shown above, the clockmaker's depthing tool is an expensive tool but is the best way to ensure all the wheels and pinions run at the best possible depth. The better the mesh, the better the clock will run. A clock movement will not tolerate a poorly set depth-mesh. (The correct term 'depth' and 'depthing' will be used from now on rather than mesh or meshing.) Using the depthing tool is covered later in the text.

The author acknowledges that the depthing tool is a task-specific tool, and the movement can be made without a depthing tool if necessary. The depthing tool is also a useful tool for marking out the wheels to 'cross' them, though if the author were purchasing one specific extra hand tool other than a pivot file/burnisher to help make this clock movement, it would be the depthing tool.

Depthing the movement without the use of a depthing tool

Alternatively, a trial set of holes can be drilled into a spare brass plate to try the depthing, although this can be a rather laborious process. This movement has been designed to allow the process of hole drilling to be as effective as possible for trying the depth of the wheels running together by deliberately placing the wheels and pinions at the ends of the arbor. The drawing of the movement gives the centre distances for the actual movement made to illustrate the project, but unless the wheels and pinions you have made are exactly the same size, the ideal centre distance *will* be different to that shown. However, these dimensions will give you a good place to start from.

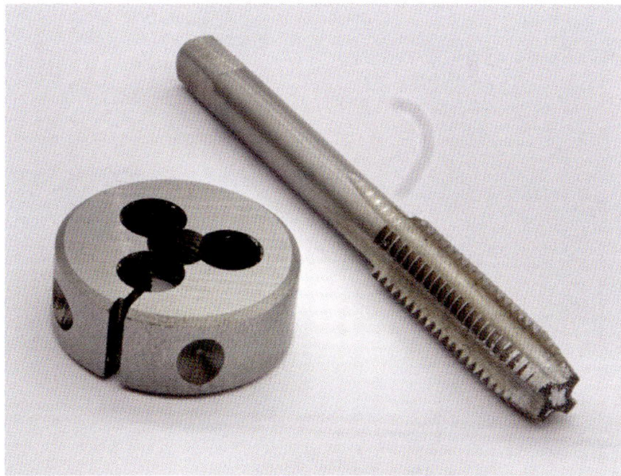

Left is a split die, typical of the type commonly used. Right is a tap, in this instance a 'second' tap.

Taps and Dies

For making screws, a set of taps and dies is required. A 'tap' is the tool used to put a thread into a hole. A set of the relevant taps will be needed. Taps are available in a number of different specifications and thread 'pitches'. For this project use either metric (mm) or BA (British Association) threads. In practice a clockmaker would use both, but for the purpose of making this clock we will focus on using BA threads, as there is a ready source of un-plated slotted steel screws available for those who do not wish to make all their own screws.

A set of taps consists of three taps, known as a taper, middle and plug. Initially the taper tap is used, care being needed to ensure it is worked through the material at 90 degrees to the surface. Next, the middle tap is passed through, and then finally the plug tap. A tap holder is the best way to hold a tap, although a pin vice can be used for the smallest taps. A table of metric alternatives (note they are not an exactly equal size) is given in the appendix.

The three variations of a tap as described.

pulley screw, so 2BA and 5BA dies are required.

It is not necessary to purchase a whole set of taps and dies, it being considerably better to buy good-quality items of the ones needed rather than a whole box of poor tools. Note a tap will not screw into its corresponding die, and will lock in place spoiling both tools.

Machine Vice for Use on the Pillar Drill

This type of vice generally has thicker jaws to ensure a parallel closing action and give a firm grip when holding things to work on with the pillar drill or milling machine if you have one. The two examples shown are a plain vice used by the author's family for approximately 100 years, and a modern adjustable angle machine vice.

Machine vice, named as such as its use is to hold the workpiece whilst being machined.

Sturdy removable bench vice with 60mm-wide jaws.

A die needs to be held in a die holder, be it in the tailstock of the lathe or at the bench. Dies are used for threading the screw in order for it to be able to screw into the previously tapped hole. If you choose to purchase screws rather than make your own, dies will not be required other than for the click screw and the

Bench Vice

This is a small but sturdy vice you can take on and off the bench. The addition of copper or brass linings for covering the jaws of the vice are required to minimise the risk of marking your work with the jaws. These are simple to make for yourself, taking about five minutes to make. The procedure to make them is as follows:

COPPER/BRASS JAW LINING FOR THE VICE.

1. These consist of a thin sheet of either brass or copper cut with tin snips to the length of the jaw vice. In the photo on the top left, the finished pair of brass soft jaws are shown being used for the filing exercise. They will be used a great deal in the project and will not look quite so pristine by the time your clock movement is finished!
2. Using a pair of tin snips, cut a length of brass the same length as the jaw of the vice, and approx 25mm wide. It does not matter that it is not terribly neat, as in operation it will get cut into and marked.
3. Place it in the vice with approximately one third above the jaws.
4. Hold a strip of wood against the brass and bend it over. Use a hammer to finish the bend to 90 degrees. Finally use a fine-cut file and round off the sharp corners to make them safer to handle.

The brass jaw lining in use, holding the exercise piece ready to draw-file the edge.

Begin with a thin piece of brass/copper; 0.8mm would be quite thick.

Place it level in the vice, with approximately one-third clear of the jaws.

Ready to bend the jaw.

BRASS FILINGS AND TURNINGS

When cutting and turning brass, filings and turnings are the waste product. By placing them in a tray and heating, these can be used for heating steel to make cosmetic changes to its colour. Just collect these as you make the various brass parts of the movement. They will be used for finishing towards the end of the project.

The hand puffer.

A hand 'puffer'

For removing small bits, filings, dust, etc., from your work without needing to blow on it. This avoids putting extra moisture onto your work.

Finishing Tools

A selection of home made 'buff sticks' can be very useful. These are approximately 250mm-long lengths of hardwood, roughly 6mm thick and 22–25mm wide, on which various grades of either abrasive papers that can be used wet and dry or finishing gel sheets are attached. An old fine-cut file makes an excellent 'buff stick'. Finishing gel finishing sheets look like abrasive paper, but are self-adhesive plastic. They give a flatter finish to the work, especially if stuck onto plate glass. Shown here are two wooden and two old file puff sticks.

Finishing papers and films

Micro finishing film

These are in fact not papers, but self-adhesive abrasive plastic films that are stuck onto plate glass. Using this system gives very good flat results when compared to regular abrasive papers. Purchase a complete starter pack unless you have a 10mm sheet of plate glass to hand. Do not cut the sheets into strips, but just in half width-wise and stick them three on each side of the plate glass sheet.

A selection of buff sticks, mounted on wood and an old smooth file.

Selection of fine finishing gel sheets mounted on a 10mm plate glass sheet.

Laying the self-adhesive gel sheets onto the plate glass.

Whilst it is not a cheap product, you will be able to make the whole clock from the one starter pack. Wet and dry papers are also necessary for finishing the surface of the clock plates. Whilst these are available in many grades the ones helpful for this project are 240, 400, 600, 800, 1200 and 2000. They are best used wet; the process is described further in the text. (*See* appendix for supplier.)

Making the Plates and Pillars

4

TO BEGIN WITH, decide which variation of the movement you are making. The layout of the clock plate is different for the deadbeat and recoil escapements. Once you decide, stay with this choice, as it is not easy to change to the alternative.

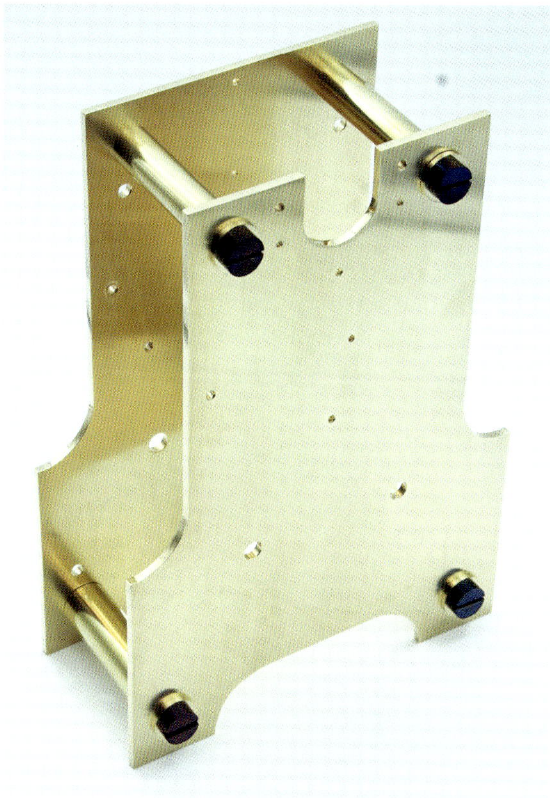

The finished plates and pillars.

MARKING-OUT TIPS

Remember, in order to have a smart-looking clock movement, all marking-out lines and any other odd marks will have to be removed during the finishing processes. Work as tidily as possible to avoid unnecessary tidying and finishing. A clock movement looks much smarter with all the surfaces flat and the corners sharp. The more finishing you have to do, the more the risk of rounding the corners and losing the sharp edges.

In order for the marking-out lines to be clear, it is best to colour the brass (and later the steel when making the pallets). Use a dark colour so that a light scribe line will show as a fine bright line through the colouring. This colouring is traditionally blue, using a form of ink known as marking blue or sometimes 'engineer's blue'. It is made by mixing methylated spirits, gentian violet and shellac. Strictly speaking, engineer's blue contains a grease that stops it setting, and can make it a little messy to use.

Marking blue is available from engineering suppliers, but two alternatives are available: purchase either a wide-tipped permanent marker, or a 'write on anything'-type of pen used to write on CDs. Colour in the material and let dry for a few minutes. For the purposes of this publication, a wide-tipped permanent marker is used.

It is helpful to begin the clock by making the set of plates and the pillars that hold them apart, to which the remainder of the movement is made and fitted. These need to be thick enough to be firm, and large enough to accommodate the mounting of all the various parts of the clock movement and dial. The actual final shape of the plates is an opportunity to personalise the design of the clock. To aid marking out and construction, it is a good idea to make rectangular plates in the first instance. Shape can be added later. Bear in mind that having a flat bottom to the plates makes mounting the movement much more straightforward. It also makes it easier to handle on the workbench. When the plates are fixed together they will stand on the bench with a flat bottom. Most weight-driven longcase clock movements are striking movements and therefore have larger plates than required for this movement. They usually have four or five brass pillars separating the plates; five in number is traditionally a sign of better quality. On quality clocks and those performing several functions, more pillars are required to securely hold the plates. Four pillars will work well for this clock, as it is a single train clock with a good thickness of plate.

The author's own preference for the shape of a clock plate is to add individuality to the basic rectangular shape. To that end, the shaped plate option is used for illustration throughout the publication. The drawings show a rectangular plate, and also a variation on the theme that has worked well for a number of the author's clocks. Both shapes of plate require the same basic dimensions of CZ120 brass sheet that is at least 3mm thick, although the location of the top pillars is different. The shaped-plate top pillars are set further in, to account for the top section of the plate being narrower than the rectangular plate. For guidance for the shaped plate *see* the section entitled 'Shaped plate variant'. Follow the dimensions shown for the plates and pillars.

MARKING OUT RECTANGULAR PLATES

When purchasing smaller sections of brass sheet, some of the edges are not cut completely square; that is to say, the corners are not 90 degrees, and the sides are not parallel. In addition it is often cut with a guillotine, which causes the edges of the plate to be slightly rounded over. None of these issues is helpful to the clockmaker, and have to be overcome.

Looking at the brass plate in front of you, which is the cleanest and straightest edge? If possible, bearing in mind the dimensions for the clock plates, use this as the bottom edge of the plate from which to mark out all the other lines. Does this edge appear slightly curved over? If so, your marking-out will need to be beyond this edge, so measure in enough to clear this edge, probably approx 3mm, and draw a straight line across the sheet, parallel to the edge.

Marking out the plates on 3mm-thick brass sheet

1. Begin by colouring the surface to be marked with blue colour, allowing it to dry.

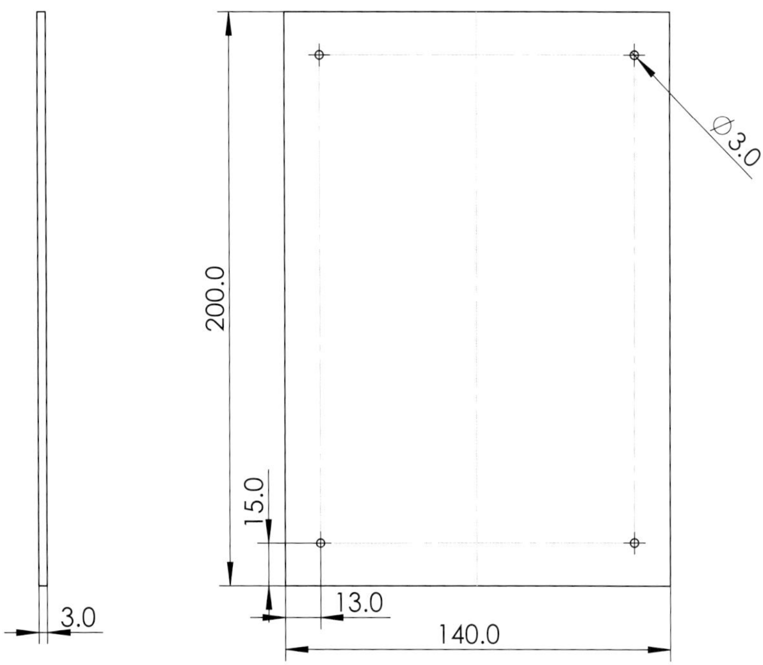

Rectangular plate.

Colour the surface of the brass to help see your marking-out lines.

The blue colour has been exaggerated in this photograph to help the lines to show up. As can be seen, they are easier to follow when the brass line is contrasted with the darker blue added colour.

2. Using a scribe and engineer's square, mark a line up the left-hand side of the brass sheet for a height of 200mm from the base – the square edge – of the sheet.
3. Now mark from the left hand edge 140mm along the bottom true edge.
4. Next mark 1mm only, and then another 140mm.
5. Using the engineer's square again, scribe up 200mm from these marks.
6. This gives two marked-out rectangles of brass measuring 200mm by 140mm and 3mm thick. The 1mm line between the plates is the allowance for sawing with the piercing saw.
7. On both plates mark a centre line from bottom to top, by measuring in 70mm from either plate edge and using the engineer's square to scribe a line. Do not make this line too deep, as it will need to be removed later when tidying and finishing the movement.
8. Having decided to fit four pillars to the movement, now is the time to mark-out the plates to locate the pillars. Measure in from the left-hand and right-hand edges of both plates 13mm, and from the bottom of the plates 15mm up. Do the same from the top of the plate down. Where the lines meet gives the location for each pillar. You should now have two rectangular plates with four pillar locations marked, and a centre line marked down each plate. (The drawing gives the location and dimensions for these holes.)
9. Choose which plate you wish to be the front plate and mark it with a 'write on anything' permanent pen. This will *always* be the front of the front plate. On this plate, centre-punch the four pillar positions. Take care to place the centre punch exactly on the point where the lines intersect. Make a single light tap on the punch with the hammer and check your mark. At this stage, if you have missed the exact point, you can re-locate and punch it again. Once the punch mark is in the correct place, make a second firmer mark with the hammer and punch.
10. On what is now the back plate, centre-punch the top-right and bottom-left marks. Again, take care with these centre marks, as they need to line up with the front plate holes when drilled.

    There are four further holes that will need to be marked and drilled in the front plate only, for mounting the dial once you reach this stage. To avoid confusion these will be marked and drilled later in the project. The location for the dial mounting holes is the same for both shapes of plate. The dimensions are shown in the drawing for the shaped plates.

Cut around the edge of the plates with the piercing saw. Note the extra cut-out for the rear plate (left-hand side above). This is to accommodate the crutch end of the pallet arbor assembly in due course.

11. Using the piercing saw in the same way as the practice exercise, cut round the edges of the plates. The thicker brass will take more time to cut: patience and care will be rewarded.

Shaped plate variant

The same marking-out technique is used for the shaped plates, but the following sections are also scribed onto the rectangular plates to form cut-out sections. Shaping the plates in this way will require the top pillar centre points being located further towards the centre of the plate.

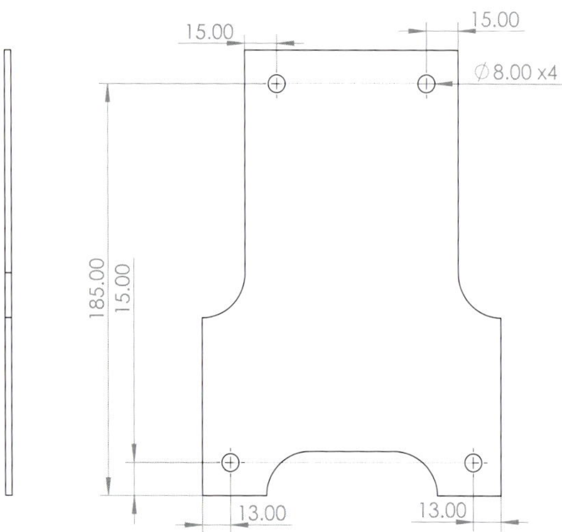

Pillar location.

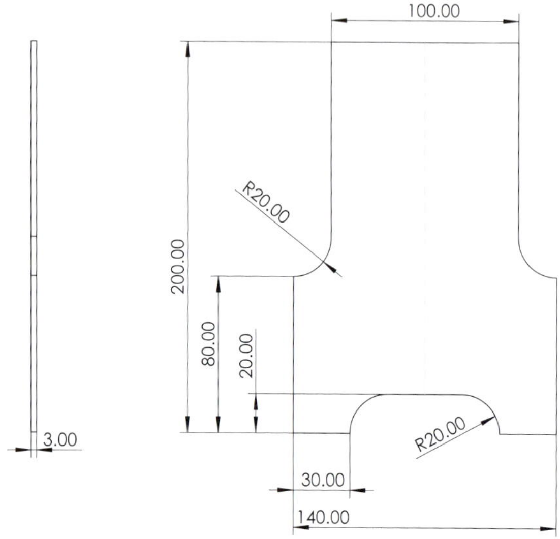

Front plate.

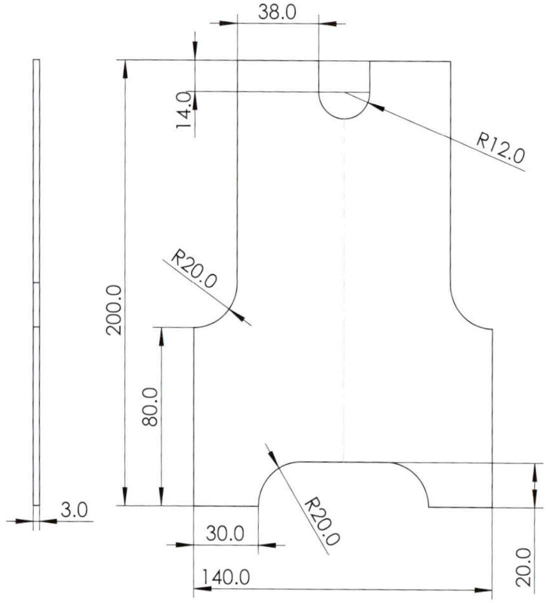

Back plate.

1. To mark the cut sections for the sides, begin by measuring up from the base line of the plate 100mm up either side. At this point make a centre punch mark.
2. Set the dividers to a radius of 20mm and, using the position centre-punched, scribe a quarter of a circle from the edge of the plate towards the middle. Imagine you are drawing from the 3 o'clock position to the 6 o'clock position of a clock dial. Do this for each side.

3. Next, measure in at the the top of the plate 20mm and scribe a line from the inside of the circles scribed to this point. This will produce a line 20mm in from the edge of what would be the rectangular plate edge, with a 20mm radius at the bottom, which gives a clean line to the lower section of plate being 20mm wider each side of the centre line.
4. To form the cut-out section across the bottom edge, measure in from either side 50mm and make a centre punch mark. Use the divider to scribe two circles, again radius 20mm, and then using the scribe join across the two circle tops. This is clearly shown on the accompanying drawing.

The back plate cut-out for both types of plate shape

There is a further cut-out required in the back plate of the movement to allow for the fitting of the escapement. This is marked out by measuring 14mm down the centre line of the back plate and scribing a circle of 12mm radius – 24mm diameter. Using the steel rule, scribe from the side of the circle a tangent line to meet the top of the plate. This will give a cut-out 24mm wide, with a smooth curved lower edge.

CUTTING THE PLATES

Place the marked-out brass sheet with one of the marked-out saw lines just off the edge of the workbench and, remembering to place the blade in the piercing saw frame to cut on the *downward* stroke, use the piercing saw to cut out the two plates. Put the blade against the edge of the brass and, taking care, guide the saw as you pull it gently downwards to cut the brass. A tip to help here – but be careful as the blade can very easily cut you – is to use your other thumb to guide these first cuts. All the hand tools can be used either left- or right-handed.

The thicker the brass used for the clock plates the longer this will take, but patience will be rewarded by an accurate cut. Many of the clocks made by the author have 5mm-thick plates, but 3mm plates will work well for this movement. Try not to force the saw blade to cut more quickly than it is designed to do. It will only cut so much at a time and further pressure will either break the blade or cause the cut to go away from your marked cutting line. Patience is the watchword here. A steady and accurate cut will take time, but will save a great deal of filing in due course. Move the brass sheet as you cut to ensure you do not cut the edge of the bench.

DRILLING THE PILLAR HOLES INTO THE PLATES

Having cut the plates the next step is to drill the holes for the pillars. Drill through the two holes centre-punched on the back plate, and through the corresponding two holes on the front plate.

In order to drill the holes accurately that pass through both plates, it is necessary to be able to clamp the two clock plates together. Traditionally this was done by riveting the plates together using pins in specifically drilled holes. Although these holes were usually filled in, on many pairs of old clock plates they can still be seen. The procedure described here is used in order to be able to screw the plates together and separate them as many times as required after an initial pinning together for the first two holes. This saves the need for continually pinning and unpinning the plates. In addition, we are using the pillar holes rather than drilling further unnecessary holes in the clock plates.

Taking the plates one at a time in the first instance, place it on a piece of flat 18mm plywood on the drill

A padded G-clamp used to hold the workpiece in position for drilling.

DRILLING BRASS

Owing to the nature of brass it can 'catch' the drill bit when it is being drilled. Care must be taken to hold the workpiece securely, and to use drill bits with the rake adjusted to zero. This is done by using an oil stone of a fine grade, called an India stone. Place the drill parallel on the surface of the stone. Draw the cutting edge back and forth a few times. It only takes a few strokes to form a flat on the cutting edge of the drill. Hold the workpiece securely to the drill table with a G-clamp; this just ensures that nothing moves when you begin to drill the hole. Whilst these drills with the cutting angle flattened will cut other metals, it is best to keep them just for brass, as non-adjusted drills will cut steel more effectively. Other than when using a centre drill, it is best to drill brass with as slow a speed as possible, both with the pillar drill and when drilling a hole in brass in the lathe.

Centre drills are shaped differently and cut right at the centre point. When using a centre drill, drill in far enough to give a good lead to the drill bit you use next. Do not go too far as the drill will grab. Please remember to wear safety glasses for all machining processes.

The drill being adjusted for drilling brass.

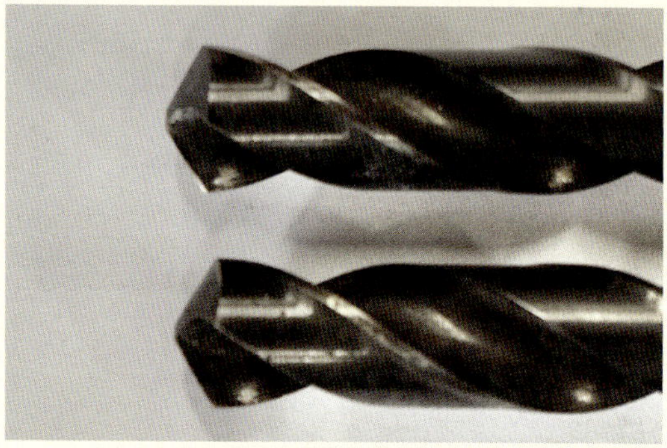

Comparing the adjusted drill – top – to the original drill.

table. The plywood board placed on the drill table is sacrificial and helps avoid drilling into the table of the pillar drill. Using a 1.5mm drill (or a small centre drill with a similar diameter) placed in the drill chuck, put the brass plate under the drill so the drill lines up with the centre punch mark. Now clamp the plate in place with a G-clamp using a packing material between the brass plate and the clamp to avoid marking the plate.

Drill through the top-left and bottom-right holes of both plates with the same drill. Put the plates together as a pair. To form the rivet use two of the pins filed in the earlier exercise that turned out tapered and use them to rivet the plates together for further drilling. Do not overdo the riveting as it will be removed as soon as the two remaining holes are drilled to the correct diameter. Push a pin through the holes in both plates until it stops, then tap it once with a hammer. Cut off both protruding ends of the pin so there is just 1mm sticking out from either side of the pair of plates. Place the plates on the anvil and hammer the ends of the pins until the plates feel firmly held together. Try not to catch the plate surface with the hammer, as hammer dents can be time-consuming to remove. There is no need to over-tighten the riveting.

Line up the centre-punched hole with the 1.5mm drill and secure in place using a G-clamp or similar device. Note the thick tape round the foot of the clamp; this is to stop the clamp marking the brass. A wooden packer could also be used.

Place the plates to one side and make the following tool to allow the plates to be easily secured together for various tasks for the remainder of the project.

MAKING FOUR TURNED STEPPED REMOVABLE BRASS PLUGS

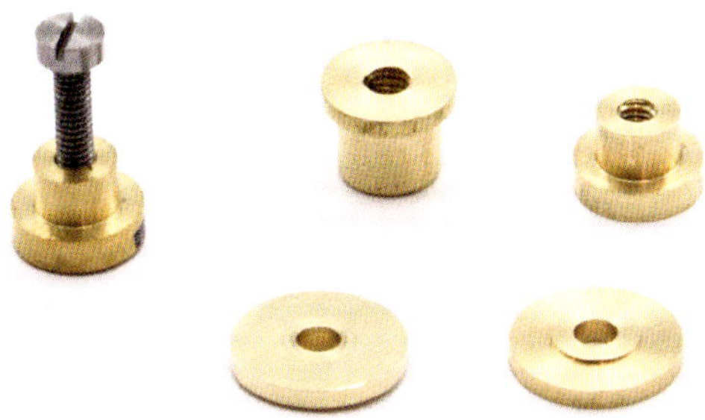

The finished plugs.

is to use a collet wherever possible, and he has the luxury of having a lathe set up for collets and another with a three-jaw self-centring chuck. Using a collet is quicker and the material is more likely to run 'true'. In addition, there is much less risk of marking the material when it is held in a collet. If you are using a chuck, it is worthwhile using a thin brass packer between the jaws of the chuck and the workpiece to stop the jaws of the chuck from unduly marking the bar.

Returning to making the plugs

Place the brass bar in the lathe with about 30mm protruding from the collet or chuck. Start the headstock turning and use a turning tool to 'face' the end of the bar. That is to say, make it flat and true across the surface of the end of the bar. This is shown in the photos

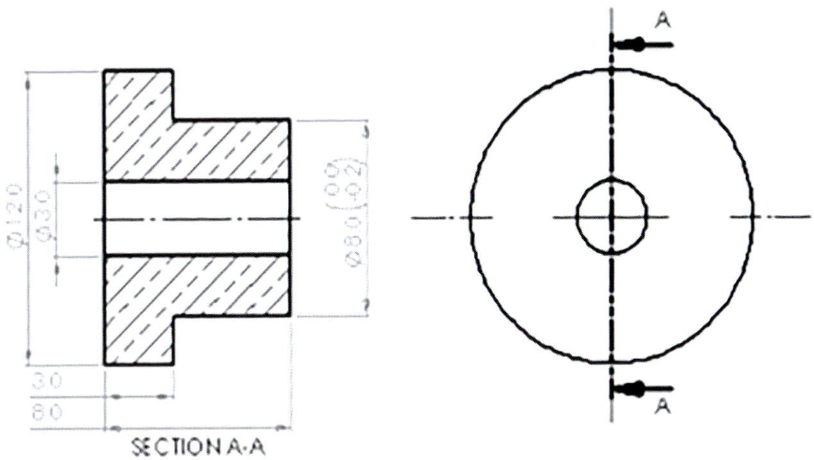

Pillar plug number 1.

Make the first simple tool to help: tool number one

When making a clock there is often quite a bit of sawing and filing, so it is good to have a change of task to keep the project interesting. Putting the plates to one side and using the lathe, turn four stepped plugs to the dimensions shown and thread them using a 4BA tap. These brass plugs are 'tools' to help make the clock, and are not used in the final clock.

They are used in place of riveting and re-riveting the clock plates together.

Use a length of brass of diameter 12mm; place this into the head stock of the lathe, either using a collet or three-jaw self-centring chuck. The author's preference

on the following page with the explanation of setting the cutter at the correct height if adjustment is needed.

Setting the turning tool to the correct height

Note: The author recommends using the type of tool holder that is a quick-release type arrangement. It makes changing the cutting tools much quicker and avoids having to carry out the centring exercise discussed every time the tool is changed.

When using a turning tool, we usually require it to cut across the centre of the work. If the cutter is high it will leave a section unturned in the middle and may leave a surface that looks 'chattered', that is, very rough. Also the cutter will not go into the centre of the workpiece.

When the tool is set too high it cannot pass over the centre of the bar, so it is not possible to face the surface right to the centre of the bar. Note also it is producing quite a coarse finish on the 131 brass bar used here. The author tends to use 131 brass bar, as the colour is a little more yellow.

Adjust the height of the turning tool in its holder to obtain a clean, smooth cut across the centre of the workpiece.

If the cutter is set low it will probably turn reasonably, but will still leave an unturned 'nipple' on the end of the bar.

It is worth taking the time to adjust the turning tool correctly. Once adjusted, lock it into place and it will be correct for the whole project. Next, change the turning tool for the parting-off tool. This is used to cut the workpiece from the bar stock in the lathe, leaving a smooth

A NOTE ON TIPPED TOOLS

There are a number of different-shaped turning tools, made from a variety of materials for various different tasks. It is a skill to sharpen a tool steel cutter to achieve the correct results. Whilst the author uses a combination of tipped and tool-steel tools, it is recommended that tipped tools be used for this project if you are new to turning. A 'tipped' tool consists of a bar capable of holding a very hard insert in the end. There are a number of cheap options for tipped tools. The author has tried a number for this publication and has found them best avoided. Use a good quality brand such as Sandvic.

Left: A tipped tool holder with a tip attached. In the middle is an unmounted tool tip. These are not re-sharpened; when worn blunt the tip is replaced. Right: A cutting tool made from tool steel and not tipped. These are sharpened using a grinding wheel.

finish. A hacksaw or piercing saw is not used, as these can produce fine swarf, which could cause problems with the lathe. A saw is also not used as there is the risk of slipping and damaging the lathe or yourself as well as leaving a poor finish, which will require the removal of metal to tidy. As we are working to 0.1mm with this project, there is not enough spare material to have to turn the end to remove the unevenness caused by sawing.

Now use the parting tool and repeat the centring exercise and lock this tool in place. Whenever you fit a new turning tool into the tool holder it will need to be centred as described. Both tools are now set and readily exchanged for the various following tasks.

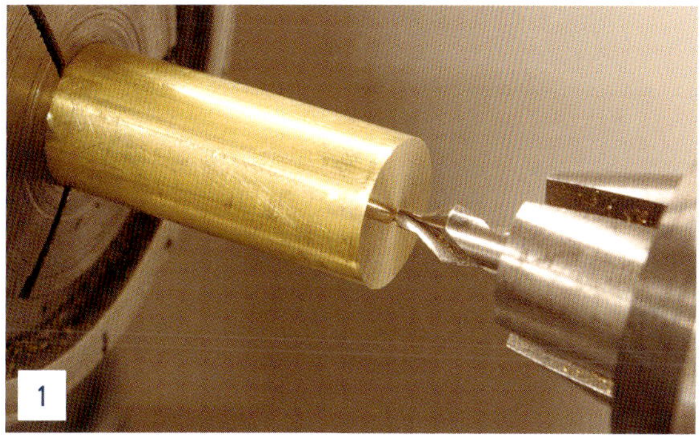

The first operation is to drill a hole in the centre of the bar. This is done using a centre drill held in the tailstock of the lathe. For this project a Jacobs chuck is used for holding drills and taps in the tailstock.

DRILLING THE WORKPIECE

In order to drill a hole in the end of the bar, actually in the centre of the bar, it is necessary to use a centre drill. These are a different shape to a regular twist drill, are often double-ended, and are specifically for centring a hole. They are usually purchased in a set of four or five, rather than to the exact diameter of the hole you wish to drill. Use a centring drill smaller than the finished diameter you require.

Place the centre drill in the tailstock of the lathe. Whilst it is preferable to hold the workpiece in the headstock in a collet, a Jacobs three-jaw self-centring chuck is generally used in the tailstock. This is the type of chuck used in the pillar drill, and in fact in most devices designed to drill holes.

Start the lathe headstock rotating and gently introduce the drill to the workpiece (the brass bar).

It will begin to cut into the surface. As we are making four plugs, it does not matter exactly how far the centre drill goes into the workpiece, provided you drill far enough for a small slope to be formed in the end of the workpiece. (*See* photo.)

Withdraw the centre drill whilst the headstock is still turning. The centre drill is then removed and replaced with a drill of correct diameter, to allow a thread to be tapped into the hole for a 4BA screw to fit – 3mm. Drill the hole 40mm into the brass, keeping the lathe running whilst removing the drill from the workpiece.

This is the procedure for drilling all holes in brass in the lathe. For safety reasons *stop* the lathe whenever you alter anything you are doing. For example, when changing the cutting tool, or a tool from the tailstock chuck.

With the headstock of the lathe turning at a slow speed, feed the drill into the end of the workpiece. Here the work is being rotated at 270rpm. Drill until the angled part of the drill begins to cut, and then withdraw the drill before stopping the lathe.

Exchange the centre drill for a 3mm-diameter drill bit, with the cutting angle adjusted for drilling brass. Drill into the brass for approximately 40mm.

Begin by turning the lathe off to avoid an accidental starting of the head stock. Place the taper 4BA tap in the Jacobs chuck and bring it to the drilled hole.

Using a Tap

The procedure for cutting a thread into a hole is carried out using a 'tap'.

As with all cutting tools, better-quality taps are worth purchasing. It is necessary to drill a hole of the correct diameter into the workpiece in order for the tap to form the thread. A very useful engineering handbook known as *The Zeus* has many useful statistics for general workshop use, including a list of which diameter hole to drill in order to correctly tap – or thread – the hole. The correct size for drilling all holes for tapping in this project are listed in the Appendix. In this instance, the diameter of the hole for tapping 4BA is 3mm.

Now place the taper tap into the Jacobs chuck – or, if you have one, a tap holder – of the lathe tailstock, and with the lathe motor turned OFF, introduce the tap to the hole, carefully turning the headstock by hand. Push the tap into the hole and whilst applying gentle pressure turn the headstock by hand one turn. This will start to feed the tap into the hole, cutting the thread as it goes. Use the action of one turn forward, working the tap in, and half a turn backwards, which will enable the tap to clear and give a generally better quality of cut thread. It also helps the tap not to stick. Use the lubricants noted. (When drilling a hole the tailstock is locked onto the bed of the lathe. When tapping a hole, **do not** lock the tailstock in position.)

Using the technique of one turn forward and half a turn back, a thread is readily cut into the workpiece. Incidentally, whether cutting a thread in the lathe

Bring the parting tool tip to the surface of the brass and, with the lathe running at a moderate speed, feed the parting tool into the bar surface. It will cut, leaving a reasonably smooth surface on both ends of the cut.

Remove the turned plug from the bar stock with the parting tool.

or the workbench, use this technique to cut a good quality thread. In addition, lubrication on the tap will help the cut. In the case of brass, traditionally a clockmaker was taught to use his own spit, which in fact works well! Modern lubrication for brass is 'Burr lubricant wax', which is a hard wax product available from UK horological suppliers. White spirit can be used, but do cover the bed of the lathe as it will wash away any lubrication you have on the lathe bed.

Having used the taper tap, repeat the operation with the second and plug taps.

Note the tap will not be long enough to thread the whole depth of the hole. Do not use the tap any deeper than the threaded section as it will clog, which may cause it to snag. As each plug is parted from the bar stock, the tap can be re-introduced to tap the thread in the next section.

Depending on the width of the parting tool being used, it may be necessary to re-drill the hole if the bottom of the hole is reached before all four plugs are made.

Having tapped the hole in the workpiece we now need to finish turning each plug to shape and then cut – part – each one from the remainder of the bar.

Using the turning tool in the compound (cross) slide, turn a step onto the end of the bar 5.0mm long of diameter 8.0mm. Next, change the turning tool for the parting tool.

Using the vernier gauge set to 8.0mm and the lathe turned OFF, bring the parting tool to line up with the vernier. Start the lathe and gently introduce the parting tool. Provided the height is correctly set, it will make a clean cut through the brass and the first plug will have been produced. Change back to the normal cutting tool and lightly face the end of the workpiece. Turn a step of 5.0mm long of diameter 8.0mm as before. Part this one off, and repeat these steps twice more to produce the four required plugs.

Making washers for the pillar screws

Having made the four plugs, we now need four matching brass washers to sit on top of the front plate. These are to ensure the screws do not mark the surface of the brass when tightening. Make eight stepped top washers to the dimensions shown. These will be used in the final clock so eight are required, one for each end of each pillar. This time use a drill allowing clearance for the 4BA screw threads to pass through; again, this figure is in the *Zeus* guide and the Appendix – 3.7mm.

Proceed as with the plugs in regard to turning, drilling and parting each stepped washer from the bar stock in the headstock of the lathe. Use the dimensions shown.

It is now possible to use the plugs and washers to secure the plates accurately together for the following procedures.

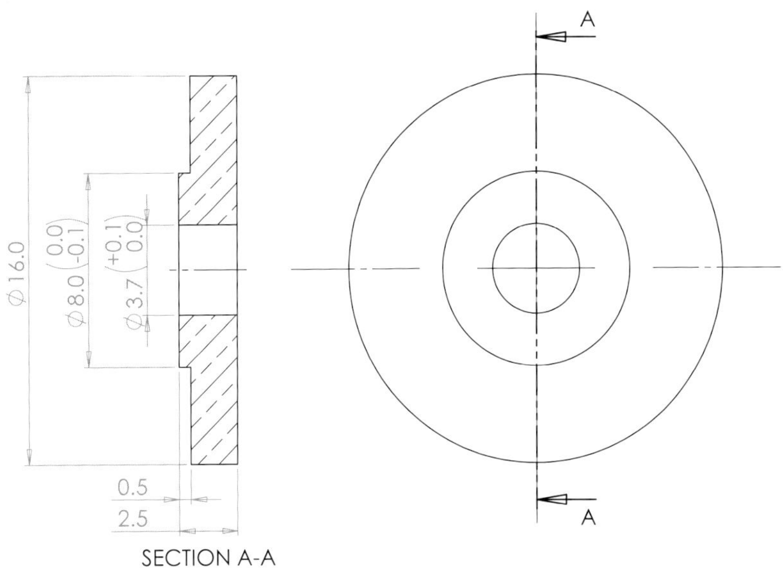

Pillar washer.

Using the plugs and washers

At this stage there are two 1.5mm holes passing through the plates where we wish to mount the pillars. These have been used to rivet the plates together.

The next procedure is to drill the remaining two un-drilled pillar holes all the way to 8mm diameter. Following this, the diameter of the two holes used for riveting the plates together is opened to 8mm – to fit the shoulder, which will be turned onto each end of the pillars in due course.

Line the tip of the drill up with the centre-punched hole of an undrilled hole.

Drill until the angled part of the centre drill begins to cut the surface of the plate.

The centre-drilled hole.

Place the plates back on the pillar drill plywood pad and, using a centre drill of approximately 3mm-diameter at the tip, line the tip of the drill up to the centre-punched hole of one of the un-drilled holes.

Note: A typical pillar drill table has slots cut to allow items to be secured to it, and for drills and the like to pass through, without damage to either the drill or table. Line up the hole to be drilled with one of these holes or slots so that the drill can pass through the table. This will allow the hole to be finished with a parallel reamer if you have one to ensure it is true.

Drill into the plate until the angled part of the centre drill is engaging with the plate. This will form a chamfer to guide the drill centrally. Change the drill for one of 4.0mm diameter, and taking care and a low drill speed, slowly feed the drill right through both plates and the plywood board underneath. Use a 4mm, 6mm and – depending on whether you have an 8mm diameter reamer – an 8mm drill bit, in order and in the same fashion. If you have an 8mm fluted engineer's reamer, drill the last hole 7.8mm in diameter and use the 8mm reamer to finally open the hole. A reamer will produce a more circular and dimensionally accurate hole. Be sure to have clearance below the plate to pass the reamer through. When using the reamer, pass the reamer through the plates and turn the drill off before removing it from the workpiece. Reaming is a low-speed operation.

Repeat this procedure with the remaining vacant hole. Remember: whenever drilling brass use adjusted drill tips. You will now have both plates riveted together with two 8mm-diameter holes drilled through two of the pillar positions.

The two plugs fitted, and the two steel rivets removed ready to open the 1.5mm diameter holes to 8mm diameter.

View of the rear of the plates shown held together with four plugs ready for filing to shape. Here the washers can be seen.

View of the front of the plates shown held together with four plugs ready for filing to shape.

Using two of the 4BA tapped brass plugs and washers previously made, secure the plates together using these two 8mm diameter holes and two 4BA screws. Use purchased 4BA screws here as they are not the final screws that will be used to assemble the clock. Next, remove the two steel rivets by filing the end off the rivet that was the thinner end of the tapered pin. A punch is then used to drive the pin out.

To finish the hole to the correct diameter it is best to use a reamer of 8mm diameter if you have one. The reamer has to pass right through the hole. The action is to pass the reamer into the hole and then stop the drill rotating before withdrawing the reamer. N.B. Ensure there is a hole through the drill table below the hole in the plate to allow the reamer to pass through.

To drill the remaining two holes repeat the process, drilling and reaming the remaining two pillar holes to 8mm diameter. The plates now have four pillar holes of 8mm diameter, which line up ready to accept the pillars once made.

SHAPING AND FINISHING THE PLATE EDGES

Whether making the rectangular or shaped plates, the process for tidying and finishing the edges is the same.

The next step is to shape the plates so they are the same and tidy the edges to make them easier to handle. Secure the plates using all four of the brass plugs made previously. Shape and then tidy the edges of the plates together. To obtain a clean file stroke, and to avoid a rather unpleasant sound, place the plates as far down in the vice as possible. (This operation can be

The two clock plates secured together with the four plugs, mounted on the edge of the workbench ready for shaping.

carried out in a milling machine if you have one. The author assumes that if a milling machine is used, the operator will be familiar with its usage.)

The author finds a typical clockmaker's bench vice too small for this, and so clamps the plates to the edge of the workbench using G-clamps with a wooden packer. Care is required when filing to ensure that the file is flat to avoid a chamfered edge to the plates. Use the engineer's square to frequently check your filing so that a correction can be made whilst there is sufficient brass left if a chamfer is occurring. To shape the plates, the file is pushed across the plate. Once the shape is reached the filing is initially tidied by 'draw filing' the edges. Place the file on the edge of the plate, and hold it at each end of the file to draw it along the edge of the plate. Take care not to knock or slide off the corner of the brass when the corner is reached, as doing so will remove the neatness of a sharp corner.

The plates will be more comfortable to handle with the edges tidied to a finer level. With the plates still clamped together use a buff stick of 240 grit and then one of 400 and 600 grit drawn lengthwise along the plate to smooth the edges. Use the same technique as draw filing.

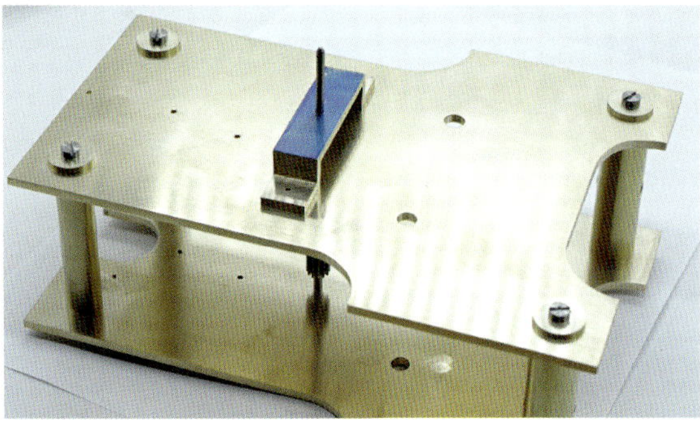

The clock movement a little further on in construction and showing the edges of the plates. It can be seen that they are much more pleasant to handle with the edges finished.

Separate the plates and there should be two identical-shaped and tidied plates. There might be a slight burr on the edge of the plates; remove this by carefully running the 600 grit buff stick along the edge. Try not to lose the 'edge' whilst removing the burr.

Further finishing of the plates will be required once the movement construction work is complete.

MAKING THE PILLARS

Making and fitting the pillars will enable the clock plates to be secured together and stand up on the bench. It is the first time your work will begin to look like a clock movement.

The next task is therefore to make the pillars to the dimensions shown. There are two readily available types of brass bar numbered 121 and 131. 121 is slightly easier to work and may be more easily sourced than 131, which is less free cutting, but is more yellow in colour. The author has used 131 for pillars and wheel collets. Whichever brass you use, the following directions apply.

If your lathe allows, use a collet in preference to the usual three-jaw chuck. When using a three-jaw chuck, take care not to mark the brass with the jaws of the chuck. Thin strips of brass or aluminium can be used as 'soft' jaw inserts if using a three-jaw chuck.

The shape of the pillars is another opportunity to personalise the design of the clock. The dimensions shown are for plain brass pillars, but provided the length between the shoulders is as shown, and the diameter of each end of the pillar correct to pass through the plate – 8.0mm – the rest of the pillar can be turned to be as ornamental as you wish. Take care not to thin the diameter too much – especially the bottom two pillars, as these pillars will be drilled and tapped to allow for mounting bolts to be secured through them. All four pillars look better if they are of the same style. The bottom two pillars are a thicker diameter in order to hold the mounting bolts.

Traditionally the pillars are permanently riveted to the back plate, and taper pins used to secure the front plate to the pillars. When making a clock today the ready supply of well-made steel screws, and the tools to make screws, means the pillars can be screwed to the front and back plate rather than riveted. This avoids the need to remove any

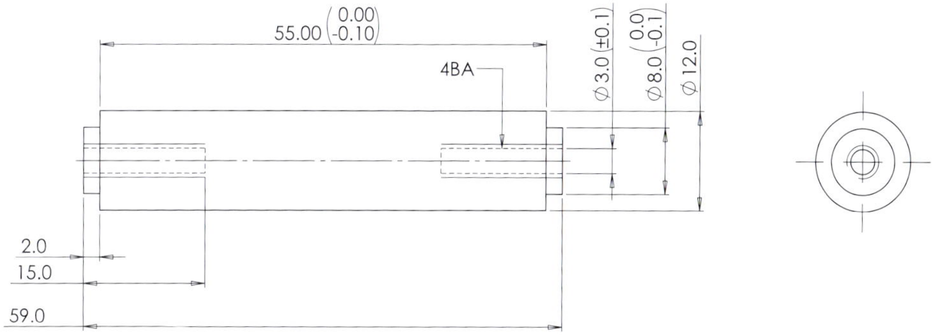

Top pillar.

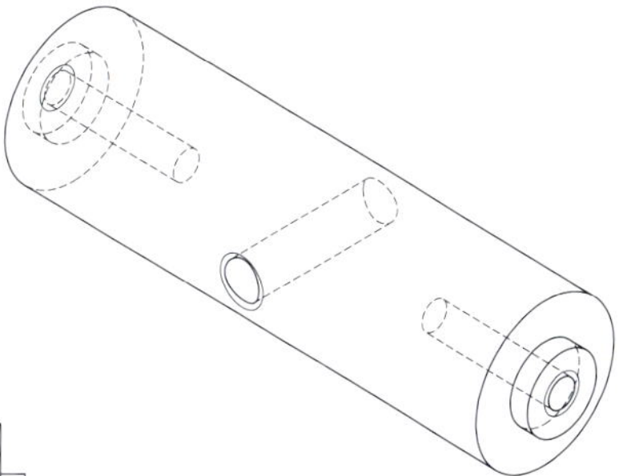

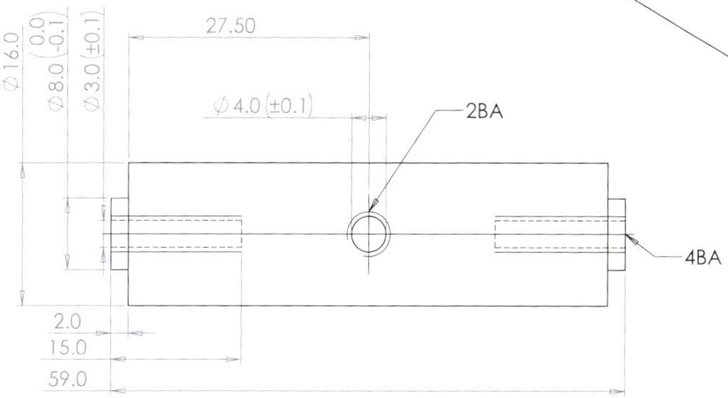

Bottom pillar.

A short length of bar sticking out of the collet allows for an accurate hole to be drilled. Drill into the workpiece until the tailstock scale shows a depth of 20mm or the tape used to mark the depth is level with the surface of the bar. Using a tape marker is not accurate to more than 0.5mm, but is sufficient as a depth marker for drilling the pillars.

riveting marks, and also allows the pillars to be readily removed whilst making and finally finishing the clock.

The process for manufacturing the pillars is similar to making the plugs, other than drilling to a specific depth of hole. Either use the scale on the tailstock, or place a piece of masking tape around the drill 20mm from the cutting edge of the drill, as shown.

Mount a piece of 12mm-diameter brass bar into the lathe with a length of 10mm being left clear of the collet or chuck and face off the end of the bar so it is flat, ready to drill.

Use a centre drill in the tailstock (tip diameter less than 3mm) to locate the centre of the pillar, then drill a hole of the correct diameter (3mm) for a 4BA tap, to a depth of 20mm. Now turn a step onto the end of the pillar as shown in the drawing.

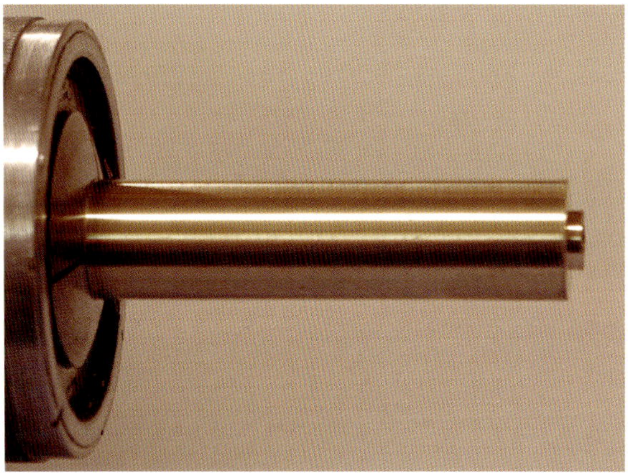

A general finish is placed on the pillars before being parted from the bar stock using 600 grit wet and dry paper.

The pillar being parted from the bar stock.

Whilst the pillar is mounted in the lathe, tap the thread into the hole. Mount the taper 4BA tap in the tailstock of the lathe, and turning the headstock by hand, one turn forward and half a turn back, cut the thread down into the hole. Use a hand puffer to blow the swarf from the hole. Now switch to the plug tap and repeat the above in order to extend the thread further down the hole. Next move the brass bar – the workpiece – along in the lathe until 65mm is clear from the collet or chuck. A generally better finish is now applied to the pillar using 600 grade wet and dry paper or equivalent of gel sheet made into a buff stick. Cover the lathe bed before carrying out the operation, as it is important not to get any debris from the abrasive sheets onto the bed of the lathe.

It will look as in the photo above left.

Using the parting tool, cut off a total length of 59mm. Place this length to one side and face the end of the brass bar before repeating these steps so you will have two pillars, both having a tapped hole, a step, and an overall length of 59mm.

Repeat this using a piece of brass bar of a diameter of 15mm to make the bottom two larger diameter pillars. The increased diameter is to allow for mounting the movement in due course. *See* drawing on the previous page.

Take each pillar in turn and treat the unworked ends in the same way as above. Thus there will be four pillars, each with a 4BA tapped hole in each end and a step turned as the drawing shows. Take care when turning the step in this end as the distance between the steps on all four pillars *must* be the same at 55mm. The dimension of 55mm between the shoulder of the pillars is important, so if you end up short make another pillar.

Drilling the mounting holes

One further operation is to drill a hole halfway along the side of the larger-diameter bottom pillars. These will be used to mount the movement. It is not easy to get the central point exactly where you want it when lining up the drill. It is also tricky to get the drill to pass through the centre of the pillar, rather than ending up slightly off-centre. There are a number of different ways this can be achieved. The author uses the following technique, which is not the invention of the author, but one which gives satisfactory and repeatable results.

Take the bar stock used to make the larger-diameter lower pillars and place it back in the lathe. (The outside diameter needs to be the same as the two bottom pillars. So if you turned down some stock to the correct diameter for the bottom pillars, turn this section of brass to the same diameter.)

Drill a hole approximately 25mm deep into the brass of 4mm diameter, using the centre drill, as described above. Part this from the lathe 20mm long. This will give a brass tube of 20mm length, diameter 15mm and with a 4mm diameter hole all the way through.

Place one of the bottom pillars in the machine vice with a packing block below and place the 20mm-long tube end on top of the pillar. See photo on the following page.

Using the vernier gauge between the shoulders of each end of the pillar, set the tube so it is an equal distance from both ends. Carefully tighten up the vice and double-check that the tube is upright and still equidistant from each shoulder. Line up the tube under the 4mm drill placed in the pillar drill and pass

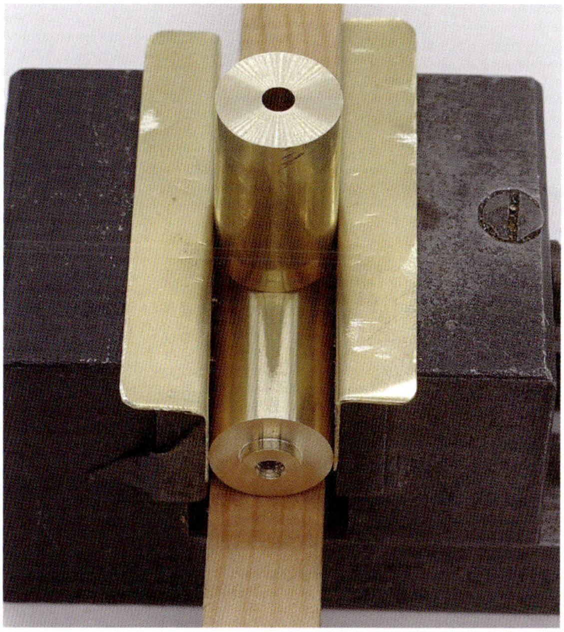

The bottom pillars with a diameter of 15mm are drilled to accommodate the mounting bolts.

Pass the 4mm-diameter drill down through the guide tube, and right through the 15mm pillar.

the drill down the tube, drilling right through the pillar. You now have a hole passing straight through the pillar at the mid-point between the shoulders. Repeat for the remaining larger-diameter pillar.

This hole will enable a 2BA tap to be used to cut a thread through it. Place the pillar in the vice, using the brass soft jaws to avoid marking it, hold the 2BA taper tap in a tap wrench and work it through the hole, using the same forward and back motion as used in tapping the pillar securing threads in the end of the pillars. Take care to keep the tap upright as it passes through the hole. Repeat, using the plug tap to give a good clean constant thread through the pillar. Later in the project two mounting bolts will be made to fit these holes.

It is now possible to place a pillar through each hole, and using one of the eight turned washers made earlier and a 4BA steel screw, assemble the movement. The workpiece will now begin to look like a clock movement.

Dial mounting holes

Now is a good time to mark-out and drill the four previously mentioned holes for the dial pillars to pass through the front plate.

Taking the front plate ONLY – we do not want duplicate holes in the back plate – mark 12mm in from both edges and 30mm down from the top; also, mark up 30mm from the bottom of the plate. Where the scribed lines meet gives us the points for the centre of the four dial pillars. Centre punch and drill these four holes (as the previous plate holes were drilled) to a diameter of 4mm.

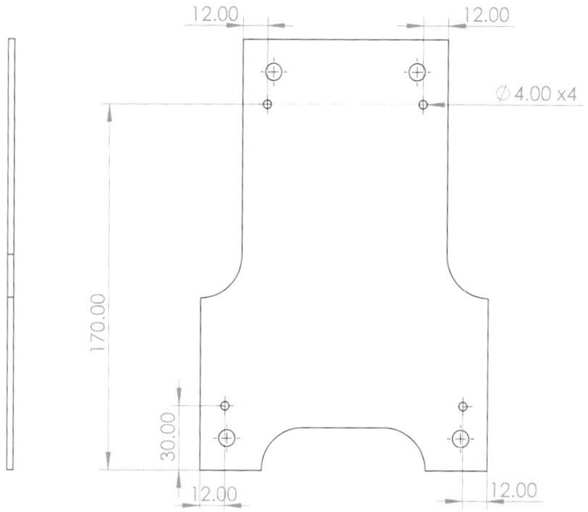

Front plate location of dial mounting holes.

5 Screws

MOST ANTIQUE CLOCK MOVEMENTS are held together with tapered steel pins. Once mass production was really underway, in America and Germany especially, nuts were used onto threaded pillars, the ends of which were extended through the plate and a thread cut on the extension to enable the nut to be fitted. This was also the practice for UK manufacturers at the turn of the nineteenth to twentieth century, with F.W. Elliott continuing to use this design up to the end of production in 1998. For making this clock, slot-headed screws will be used. Cross-head, Allen head, or pozidrive heads are generally not used in clock making for aesthetic reasons – you may differ in your view on this point, though. The author has used Torx head screws for one project.

As the screws are going into brass, use steel screws, as they will grip well and will be less likely to bind in the threaded holes.

Steel slot-headed screws of both metric and BA threads are readily available. You can also buy zinc-plated screws, although these are not recommended for this project, as it is not so easy to finally finish the screw as it is completely plated. Although the plating stops the screw from rusting, the plating is also in the thread and slot of the head, meaning it will not heat-treat to change colour to blue for finishing. The main screws – those screwing into the pillar ends – for the finished clock look best if the head diameter is larger than commercially available screw heads. If you have a 4BA die, making these screws is easily achieved. They can then be polished and blued to add a decorative finish. The remaining screws can either be purchased or made.

MAKING THE PILLAR SCREWS

Make these screws to the dimensions shown.

Place a piece of 16mm-diameter steel bar in the lathe with 30mm showing in front of the collet.

Turn the diameter of the first 12mm to 3.6mm; this is to cut the 4BA thread onto. It is helpful to turn a small taper on the end of this section to enable the die to feed on and begin cutting the thread more smoothly. The next section of the bar should be turned to diameter 15mm. This is for the screw head and also parting the screw from the bar.

Place the die in the tailstock die holder. Die holders are inexpensive and will greatly help cutting the thread accurately. Put some regular cutting or general lubricating oil onto the steel to lubricate the die as it cuts.

In the photo a piece of white paper can be seen placed on the lathe bed. This is to aid the photograph

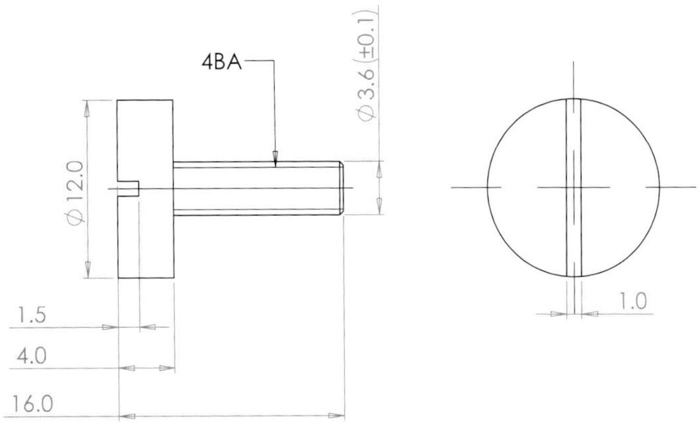

Pillar screw.

Place the steel bar in the lathe.

For the first 15mm, the diameter should be turned to 3.6mm. The next section should be turned to 15mm diameter.

Note the slight taper shown on the end of the screw to help the die feed onto the turned section to cut the thread.

Place the die in the tailstock die holder.

Use the parting tool to part the screw at the top of the head.

and also avoids an oily mess needing to be cleaned from the lathe bed.

Bring the die to the work and begin to turn the lathe headstock *by hand*. Once the die has begun to cut the thread, use the one turn forward, half turn back method.

Cut the thread up to the underside of the screw head and reverse the headstock to wind the die back down the thread to remove it.

Setting the vernier gauge to 16mm, line up the edge of the parting tool and then part the screw from the bar stock. Place this to one side, and repeat the above steps to make seven more screws. This gives eight screws for securing the pillars.

Cutting the slot in a screw head

There are a number of ways to cut the slot in the head of a screw. Whichever method you use, ensure the slot passes through the centre of the screw head.

The middle of the screw head can be seen from the pattern left by facing the end in the lathe.

The general set up for the author's Schaublin lathe.

Probably the neatest way the author has seen is to use a fine toothed slotting saw in the lathe or in the mill if you have one. Shown here is a saw of 1mm width and 50mm diameter with 80 teeth; it is referred to as a fine cutter. The slotting saw is used in the same manner as a pinion tooth cutter. This is explained fully in Chapter 8.

A narrow needle file can also be used, care being needed to ensure straight sides to the slot, and the slot being in the middle of the screw head. Place the screw with the head sitting uppermost in the vice, using brass soft jaws. Be careful to clamp the screw firmly enough to hold it in place, but not so firmly that the thread is damaged. Use the piercing saw to cut a fine slot across the centre of the screw head. This will give both clearance for the tip of the file, and also a visual aid as you introduce the file to the edge of the screw head. Work across the head of the screw, following the saw cut, passing through the centre point with your file. Whilst it takes a little time to carry out this operation it is quite satisfying to do. Lubricating the file with white spirit helps to stop the file clogging and sticking. Holding the needle file in a pin vice will also help to hold it securely.

Polishing the screw head in the lathe

Before beginning to polish the screws in the lathe, place a covering over the lathe bed to stop any polishing products from coming into contact with it. The author uses a separate old lathe for screw polishing. There are several alternative ways to polish screw heads.

Flat polished screw heads do look lovely; this is best achieved by using a lapping wheel. The majority of clocks made in the past tended to have slightly domed screw heads, rather than the very flat screw heads generally

The slotted screw head. The slot was cut with the slotting saw; it is shown placed in the lathe ready for finishing.

Steel disc, mounted on the tailstock of the lathe. The 60 grade of gel sheet is attached to the steel disc.

found in watch movements. To this end there were a number of very useful lathe accessories available to 'lap' screw heads in the watchmaker's lathe, but it is quite possible to make a lap wheel if wanted. Traditionally a steel lap would be used with fine oil-stone dust mixed with oil applied to the surface. This would be followed by a lapping wheel made with a softer metal such as zinc, used in conjunction with diamantine powder mixed with a light oil.

Whilst this does indeed give a superb polish to the screws, skill is needed in the use of a lapping wheel in conjunction with diamantine.

A good polish can also be obtained whilst still keeping the screw head reasonably flat by using the self-adhesive microfinishing film sheet. This is stuck to a steel disc and held in the tailstock of the lathe so that it can be rotated.

To begin this process, place the screw in the lathe holding the thread with the head proud of the collet. Using a high speed to achieve a good turned finish, turn the screw, removing a further 0.5mm from both the outside diameter and height of the screw head. This gives a better surface to begin the polishing process.

Note: The following grades do not refer to typical abrasive paper sheets, but to the gel sheets.

Begin by using the coarser grade marked as 60 secured to the steel disc and rotate the headstock back and forth whilst holding the steel disc with the finishing film attached against the screw head and rotating the disc by hand. This in effect is a lapping wheel.

Continue until the turning marks are removed from the surface of the screw. Wash any residue off the screw head from this operation with white spirit. Change the microfinishing film for the 30 grade and use the same technique to remove the 60 grade marking. Repeat the process, working through the range of finishing film sheets to the polishing grade of the pink film sheet and the effect as is shown. Do not forget to wash any residue off with white spirit between each grade of film sheet. Take care not to wash the residue onto the bed of the lathe.

The finish that can be readily obtained using this technique. Looking at the screw head, the thread of the collet – which is out of focus due to depth of field in the foreground – can be seen clearly reflected in the screw head.

Using buff sticks to give a slightly domed polished finish

Another method for screw-head polishing, which is quicker, and whilst not producing a flat head, will produce a smart looking screw head, is as follows.

As with the method above, the screw is held in the lathe and turned, removing a further 0.5mm from both the outside diameter and height of the screw head. Next, use a buff stick made with an old file, and either 240-grade abrasive paper or the 60-grade finishing film stuck to the file to remove the turning marks. Change to a 400-grade abrasive paper or 30-grade finishing film, to remove the 240 marks and so on until either 2,000 grade abrasive paper or the pink film sheet has been used. Now use a cream polish, such as 'Autosol' on a blank hardwood buff stick and, using a fairly light pressure, apply this to the screw head to polish it.

Finally, there is the edge of the screw head to polish. This is done by using the microfilm gel sheets secured to an old smooth file or flat piece of steel of a similar size to a hand file. Work through the grades in the same way as the top of the screw has been treated. This will result in a polished screw head that just needs washing off in white spirit, and drying with absorbent paper.

The finished plate screw.

Bluing screws

As can be seen from the various preceding photographs, although polished screws do look very smart, blued polished screws are a very neat finishing touch. The better the screw head has been polished, the better it will turn blue. Make sure the screw head is clean and grease-free before bluing.

The blue colour is a result of careful heating. There are two ways to heat the screw without adding direct heat. The first is to take a piece of brass plate 3–5mm thick and drill a hole of 4mm diameter to enable the screw thread to easily pass through. The plate used in the author's workshop has a number of holes to suit

The spirit burner heating the underside of the container holding the brass filings.

Place the screw right into the brass filings and heat the tray from below. The author's small brass dish is an old uncut clock spring barrel with a handle attached.

different-sized screws. Add a handle to the plate so you can hold it in the vice.

The second way is to use a small brass dish filled with brass filings. For the plate screws this method works better, as the larger head is heated more evenly and it avoids the possibility of the end of the thread being heated directly. This technique is described.

For both methods, the heat is applied to the underside of the brass using a gentle blowtorch or a spirit burner.

As the brass filings heat through, the heat is transferred to the screw, causing it to begin to change colour. Firstly it will appear slightly more polished, then the colour will begin to change, going first to a straw colour, then purple, and then a deep royal blue. As the blue appears, immediately remove the screw from the tray of filings and cool it on a cold brass plate to draw the heat away quickly. With a bit of practice a lovely even and matching colour can be achieved on all the screws. If the screws are heated further the colour changes to a grey and far less-attractive colour. If this occurs, the procedure is to place the screw back in the lathe and re-polish the top and side of it starting a grade before the one used to finally polish the surface and work up to polished finish again.

Tip: Do not be tempted to blue the screws by holding them directly over the flame of the spirit burner; the colour will not be so good, and they tend to rust quickly.

HOW THE COLOUR CHANGES

This is how the colour changes as the screw head is heated. These photos were taken using a screw surplus to requirements, as the slotting did not work correctly. Not everything works correctly the first time!

On heating the screw it first appears slightly more highly polished before beginning to change colour.

The first colour to appear is a straw colour, which darkens as further heat is applied.

Purple follows straw. Blue follows purple, but the heat must be removed before the colour goes beyond blue.

If too much heat is added, the colour passes the royal dark blue to a lighter less attractive and uneven blue. The edges of the slot cut in the screw head have gone through to grey on the left-hand side.

Making the Wheels

6

CLOCKS ARE VERY PRECISE in their construction – but in a different manner to how one might manufacture an internal combustion engine, for example. The wheels and pinions in the movement of a clock run without lubrication at the point the teeth mesh. This is because changing atmospheric conditions alter the viscosity of any oil used, which would cause the clock to run faster or slower. It would therefore prove to be an unpredictable timekeeper. Further issues would also arise as the oil aged. One of the many things that made John Harrison's work on his chronometers so good was the lack of need for lubrication and therefore avoiding problems it can cause in timekeeping.

In horological work we use a tooth profile known as the cycloidal form, rather than the involute form, which is the type usually found in gearboxes. The shape and design of wheel and pinion teeth is well explained in *Wheel and Pinion Cutting in Horology* by J. Malcolm Wilder.

Horological wheel work.

For this project, the correct profile for the teeth is produced by using P.P. Thornton's Successors multi-tooth cutters of the correct module for the various wheels and pinions to be cut. The various forms are illustrated in Chapter 3.

There are companies that supply ready-cut wheels to order if your machinery does not enable you to cut your own. Buying the wheel blanks with the teeth cut would save the cost of purchasing cutters if your intention is to make only one clock.

There are quite a number of other options for wheel and pinion cutting available for a variety of lathes and mills.

Shown overleaf is a Schaublin 102 instrument maker's lathe. Other than the initial cutting of the blank from the sheet of brass, and final crossing of the wheel, it is possible to make the whole wheel blank on the lathe, and so this is the method illustrated for wheel manufacture in this publication.

A Schaublin-type lathe bed, used with a Schaublin number 12 mill, a Schaublin haunched back vertical slide, an old Schaublin compound slide, along with an American manufactured 'Sharmer' CNC dividing head. This combination allows for quicker indexing when wheel cutting.

The head stock of the Schaublin 102 with the dividing head arrangement fitted, although not engaged in the image. A similar arrangement is possible with the Myford lathe amongst others. This is the method used for this publication.

A Chronos wheel-cutting tool.

Also shown is a Chronos wheel cutting tool. These are quite a cost-effective second-hand tool to acquire for wheel cutting if your lathe cannot be utilised. Chronos produced a separate pinion mill for pinion cutting.

In addition, if the crossing and mounting stages of the process prove to be problematic you will have a couple of spare wheels to hand.

HOLDING THE WHEEL BLANK IN THE LATHE

A mandrel – which is a shaft with a means of securing an object to to be turned in the lathe – is required for holding the wheel blanks in the lathe. The quality of the tooth cut is much improved by using a backing plate or cutting three blanks together. Do not try cutting the teeth onto a single blank, as without it being well supported either side the quality of the cut will be poor. A support washer can be made to fit either side of a blank disc, but it is much more straightforward to make three wheels. Use the middle of the three if you intend to make one clock. Keep the other two for spares or more clocks if you enjoy making this one!

MAKING A TOOL: THE MANDREL

From a good-quality steel, such as EN8, and of diameter around 20mm (what you may have to hand of approximately this diameter would do but, if purchasing, 25mm is sufficient) make two mandrels as illustrated. It is important that the wheel blanks are a snug fit to the mandrel. When turning steel of this diameter, use a spindle speed of approximately 1,100rpm.

The first mandrel has a turned shaft diameter of 6mm to locate the wheel blank. Use an M6 'Nyloc' nut to secure the wheel blanks. A Nyloc nut is a steel nut with a nylon insertion to prevent it coming loose. This nut must not loosen under any circumstances whilst cutting the wheels. If in doubt, use two together.

The 8mm mandrel used for this project.

Four blanks on the mandrel ready for turning to the correct diameter. In this instance four blanks are cut together to aid making the various wheels to illustrate this publication. Two locking nuts and a washer are used to firmly hold the blanks in place, as they must not rotate.

Use an M6 die in the same way as the BA thread was used to form the thread for the plate screws. (The M6 die will cut onto the 6mm diameter shaft.) The turned section for mounting the wheel should have a smooth finish, allowing the wheel blank to pass smoothly and snugly – it must be a close fit – along it.

To finish off the turning of the mandrel, turn a point onto the end of the mandrel. This will allow the Thornton's rotary wheel cutter to be centred in due course.

A centring scope can also be used mounted in the tailstock of the lathe to centre the rotary cutter if you have one. Whilst a centring scope is ideal, a point turned carefully on the end of the mandrel will suffice. You can also mount a point in the tailstock and centre the cutter with this.

The second mandrel of shaft 8mm in diameter will enable us to cut teeth onto a blank the centre hole of which will be the correct diameter for the barrel arbor. An 8mm die is used for this.

MARKING OUT FOR THE ROUGH CUTTING OF THE WHEEL BLANK

Remember: as with all marking out, the heavier your marking lines are, the longer it will take to remove them when finishing the workpiece.

Taking a sheet of brass the correct thickness (*see* wheel data table) set the divider to the radius plus 1mm for the wheel you are making. The circle scribed will therefore be 2mm larger in diameter than required;

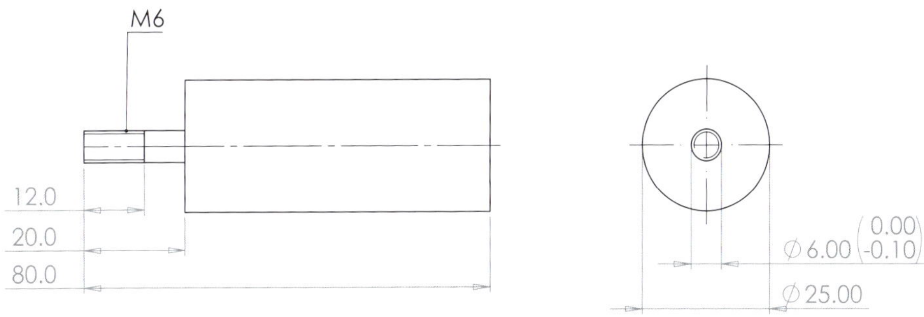

6mm mandrel.

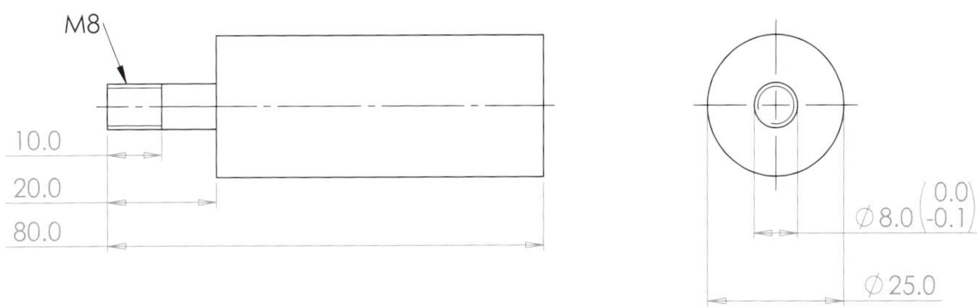

8mm mandrel.

Turning a point onto the end of the mandrel.

Colouring the brass with a blue pen helps to make the marked line much clearer, as can be seen.

this gives some margin for wayward cutting with the piercing saw. Bear in mind that brass can be purchased with one side polished. If you take care with the polished side, it will save time finishing the wheel in due course. Do any marking-out on the unfinished side.

Use the centre punch to mark the centre position, then scribe the circle set on the divider. Make three of all the wheels other than the great wheel. As the great wheel is thicker, cutting two will give a satisfactory result when cutting the teeth. Now, bearing in mind the previous observations regarding drilling brass, drill the centre hole the diameter shown – 6mm for the three train wheels and 8mm for the great and also 8mm for the maintaining power wheel if you are making this option. If you have a 6mm and 8mm reamer, use them as the centre of the hole will be better. When using a reamer drill slightly under the final diameter, 5.8mm and 7.8mm respectively here would be sufficient.

The next step is to cut the wheel blanks from the sheet using the piercing saw. It is always good practice to cut closely to the line, but do not go inside the line. Next, gently stroke the blanks across the previously prepared micro-finishing film just to remove any burrs that might have been generated.

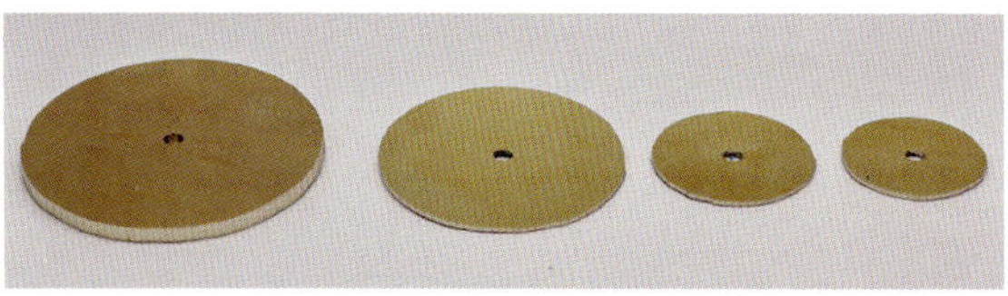

The different wheel blanks cut for this clock, left to right: great wheel; centre wheel; third wheel; and escape wheel.

MAKING THE WHEEL

Place the blanks together onto the mandrel, secured with a washer and the M6/8 Nyloc nut(s), and place the complete assembly in the lathe.

Turn the brass blanks down to the diameter required for the wheel you are making. Use the vernier gauge and check the diameter often whilst turning to ensure you do not go too far.

Next, remove the cutting tool holder assembly from the lathe saddle and place the vertical slide – with milling spindle attached – onto the saddle. Mount the relevant wheel cutter to the spindle, taking care that it will rotate in the correct direction to cut (*see* top right photo on opposite page).

Centring the cutter

The cutter needs to cut centrally across the wheel on which you wish to cut the teeth. If this is not correctly set you will cut teeth that slant, and give yourself the opportunity to do them again! It is a good idea to begin with the centre wheel, then if the tooth cutting does go wrong the diameter of the blanks could be reduced to the diameter of the third wheel whilst still mounted in the lathe. (This would save time already spent, and also the materials, although new blanks will need to be cut of the correct diameter to make the centre wheel.)

Looking straight down at the point on the mandrel holding the wheel blanks, using the cross-slide controls on the lathe saddle, bring the cutter on the milling spindle towards the point, lining them up. The use of an optical aid is helpful here. A centring scope in

Important tip:
Note when cutting the clock wheels in the lathe, do not remove the blanks from the lathe until the diameter has been turned and all the teeth cut.

the lathe tailstock is also very helpful if you have one available. The author uses an 'Optimiser' visual aid as this is rather more cost effective, and a generally useful workshop tool. Once the cutter is centred, lock the adjusting wheel in place as this adjustment must NOT be altered until after all the teeth have been cut on that wheel.

The three centre wheel blanks turned down to the correct diameter ready to 'cut the wheel' – cut the teeth into the rim of the wheel.

Dividing the wheel into correct number of teeth

Important tip:
Once you begin the process of cutting the teeth onto the wheel blank DO NOT turn the dividing head in the opposite direction to the way you are going under any circumstances. Should there be any backlash in the dividing head arrangement it will cause the divisions to be unequal. Unfortunately, this only becomes apparent as the final two teeth are cut – usually with the last

The sharp edge of the cutter teeth should rotate towards the wheel blank in an upwards direction. Therefore, in this example the cutter would rotate anti-clockwise.

tooth being too thin. When this happens the wheels are useless – the only thing to be done with them is to turn the incorrect teeth off, and use them as a blank for a smaller diameter wheel on another occasion.

Set the dividing arrangement you have to the correct number of teeth for the wheel you are making. Again, once this is set, lock any adjusting screws in place to avoid accidental adjustment.

Setting the correct number of divisions

This will depend on which tooling you have, of course. In the case of a traditional style of wheel-cutting engine – as they were called – such as the Chronos one shown earlier, you simply choose the ring of holes on the indexing plate to either match the number of teeth you wish to cut, or a number that divides equally into the number of teeth required. For example, if you require 60 teeth and there is not a ring of 60 holes, you could use a ring of 120 holes and count every other hole to index for 60 teeth.

With a worm-drive type of system mounted on the lathe headstock, the selection of which ring of holes to use is made by consulting a table. This shows the number of teeth required against which ring of holes to use. You will need the correct table for *your* dividing arrangement. The following numbers refer to the Schaublin arrangement.

In addition to the plate with the holes in – the index plate – there is a set of brass dividers mounted above

The author's Schaublin 102 set up for wheel cutting.

The dividing arrangement on the Schaublin 102 lathe. The number of teeth being cut here is 96 for the centre wheel. At the heart of this dividing tool is a worm drive at 60:1. That is to say, 60 turns of the dividing lever equals one complete turn of the lathe headstock, holding in this case a wheel blank. The Myford lathe dividing tool is also based on the 60:1 worm. There are a number of dividing plates with varying amounts of holes within them to allow for the division of many numbers.

the index plate so that when using a ring where holes are passed over there is no need to count how many holes to miss for each tooth. Set the dividers and work between the correct number of holes.

The Schaublin system for indexing in this way is similar to the Myford arrangement. Shown is the arrangement fitted to a Schaublin 102 lathe.

To give an example of setting and using this arrangement the centre wheel for this project has 96 teeth. Looking in the table accompanying the Schaublin 102 it indicates the disc with a ring of 32 holes is required to index 96 teeth.

This disc with the ring marked for 32 holes is mounted on the dividing attachment and the brass dividers placed over it. The securing nut, locking clip, and indexing lever are placed in position so that the pin of the indexing lever lines up with the correct ring of holes.

Note the tape with the arrow is a visual aid to remind the user to turn the indexing lever anticlockwise. You can work clockwise, the author is left handed so it feels more natural to turn anticlockwise. Whichever way you choose, stick to it.

Consulting the table once again, it lists the number of holes to be counted round for each tooth to be cut for 96 teeth as 20 holes. To clarify: for the Schaublin dividing arrangement, the location pin on the rotating arm of the dividing head is placed in the first hole. A further twenty holes are then counted – not counting the hole with the pin in (*see* photos on opposite page).

If using a different arrangement, check which hole in the ring of holes you should begin counting from. (The Myford dividing attachment is the same in this regard.) Set the dividers with this number. Remember *not* to change the direction of the indexing handle as you cut the teeth. Just to emphasise, you cannot change direction once you start, as any backlash in the dividing arrangement will throw the divisions off. This is annoying to say the least, as it only becomes apparent as you cut the last tooth and it turns out too thick or thin, meaning the wheel is unusable.

Setting the cutting depth

Setting the depth of the cutter is a matter of trial, error and experience. If you cut too deeply the wheels will have a reduced outside diameter, thin teeth, and may

Set the dividers and work between the correct number of holes.

not work. However, you must cut deep enough to produce the correct tooth profile or the tooth will be blunt – having a flat top – and will bind or butt when meshing in the pinion.

Cutting the teeth on a typical train wheel

Begin by setting the height of the cutter to about 1mm lower than the top of the brass wheel blank. Once again, double-check any locking screws are locked and the cutter is revolving in the correct direction.

The indexing lever moved out of the way to show the dividers set at 20 holes plus the one the indexing pin acts. The 32-hole ring is third in from the outer edge.

Start the milling spindle, running as fast as you can, but ideally at at least 4,000rpm for a good clean cut. When the author made his first clocks the milling machine he owned would only give 2,000rpm. The quality of the cut was poor, and it took time, but the wheels worked well enough in the end. The author mentions this to encourage you to go ahead and make your clock even if everything in the workshop is not ideal. There is a great deal of pleasure to be had in making a working clock movement!

To make a cut, pass the revolving cutter over the edge of the brass blanks with a smooth slow action, 'feeling' the cutter passing through the brass. Bring the cutter back through the 'cut' and make sure it is clear of the wheel before moving the index forward to cut the next tooth.

To illustrate, the first cut shown overleaf was a bit too shallow for a first cut, so another cut was taken shown on page 67. Remember, if you make a shallow cut you can always make another cut, but a cut too deep will spoil the work. (*See* for when a wheel may be used if cut slightly too deep.)

Move the wheel blank to the next position and make another cut. What you now have is one tooth cut with a slightly flat top. This shows the cutter needs to be lowered into the blank a small amount more. Every time the cutter is lowered this flat top will reduce in size until it is no longer a visible flat, and the tooth has a 'tip'.

The position as shown is set ready to cut the first tooth.

The silhouette of the great wheel teeth illustrates the shape of the tooth required.

Here the first cut taken was a bit too shallow, so another cut was taken.

It is not possible to say exactly how much further the cutter should be lowered from the initial cut to achieve a tipped tooth as each machine varies.

Remember the cutter can be lowered to make a deeper cut, but *cannot* be raised if you cut too deep. Take your time. As you cut more wheels you will get the feel of how much to lower the cutter when setting the depth.

Lower the cutter a small amount, no more than a quarter turn on the vertical compound slide and, *without* advancing the wheel blank make a further cut. Now look again at the flat top on the tooth, how does it look compared to how it was? For example, does it look as though it is half the thickness it was?

To check, move the wheel blank onto the next tooth position and make a fresh cut. On examination you will find either the top still has a flat – although it will be smaller – or there will be a 'tip'. Provided there is a flat, it will be smaller than the first time, and thus give an indication how far the cutter goes down with a quarter turn of the vertical compound slide. Repeat the steps 2 and 3 again, checking carefully each time until the flat has gone and there is a tip on the tooth.

If the cut is made too deeply this will show by the appearance of a step on the leading edge of the blank. If the step is very small the wheel may still work so do not despair just yet. Remember, the cutter cannot be raised again so lock it in position and continue to cut all the teeth.

Once the tooth has a tip, the correct depth has been found for cutting the teeth on the wheel. Lock the vertical slide in position and continue to cut all the teeth, not forgetting to go right round and through any of the first teeth that might be flat topped.

Having cut all the teeth, the cut wheels can be removed from the mandrel. If the wheel made is the great, centre, or third wheel, measure the outside diameter; provided they are within 0.5mm of the intended diameter, the wheel will work. If the wheel is 0.5mm or smaller in diameter do not proceed to cross the wheel at this point. Wait until the pinion with which the wheel will mesh has been made, then use the depthing tool to see if the wheel will run smoothly with the pinion. The motion work wheels – that is, the cannon pinion wheel, reverse minute wheel with its pinion, and hour wheel – need to be within 0.3mm in diameter to depth well enough to work. This is because there are two centre distances for the three wheels and the pinion. The depthing has to accommodate the depthing of the hour wheel and reverse minute pinion as well as the minute wheel with the cannon pinion.

The winding and maintaining power ratchet wheels are used during the process of winding the clock, so do not mesh with other wheels. There is therefore room for some extra tolerance on the diameter for these two wheels.

The above process is repeated with all the train wheels other than the escape wheel. The mandrel is changed for the great and maintaining-power wheels, using the 8mm diameter mandrel as the barrel arbor diameter is 8mm.

The cutter was lowered further and another cut made. This time the cutter is deep enough to make the edge of the tooth look rounded.

When the cutter is a fraction too deep, a small step where the 'half' of the tooth can be formed by the cutter. In this instance it is approx 0.1mm deep, which will equate to the diameter of the wheel being 0.2mm less than the correct dimension. This is small enough to be negligible for the centre wheel of this movement.

As a final note, any roughness on the tooth tip means more friction in the train of the clock, and more weight will be required to drive the clock.

Cutting the Escape Wheel

When cutting the teeth for the escape wheel quite a lot more brass is removed per cut, so the action with which you feed the cutter through the blank must be slow and as smooth as you can. It is the author's practice to cut escape wheel teeth by going round the wheel at least twice, before lowering the cutter to the final depth for the third cut. Also, if possible use at least 4,000rpm as the cutting speed for the spindle. The depth and centring of the cutter is set in a different way. The cutter is set so that the flat edge of the cutter is in line with the centre of the wheel. This is best shown by the illustration. Note that the recoil cutter is set in a different manner to the deadbeat cutter (*see* pages 73 and 74).

For this project the Brocot design of deadbeat escapement is used. The teeth for this type of deadbeat escapement are cut to be radial, and *not* offset as for the traditional English deadbeat escapement. This makes it a more straightforward wheel to cut.

The escape wheels will work with a tolerance of 0.5mm on the diameter since this can be taken up by recalculating the centre distance between the escape wheel and pallet arbor centres. With the recoil escapement a certain amount of fine-tuning can be achieved with the pallet impulse surfaces. These aspects are described when making the pallets.

The depth of the cutter is set differently when cutting the escape wheel, as the tooth has a different profile to a train wheel tooth.

The teeth cut all the way round the wheel. You need to check the first and last teeth cut are the same size. If the last tooth is a bit thinner or the first tooth a bit thicker, the wheel will not work and must be cut again on new blanks.

These drawings show the dimensions and crossing options for the escape, third, centre and great wheels.

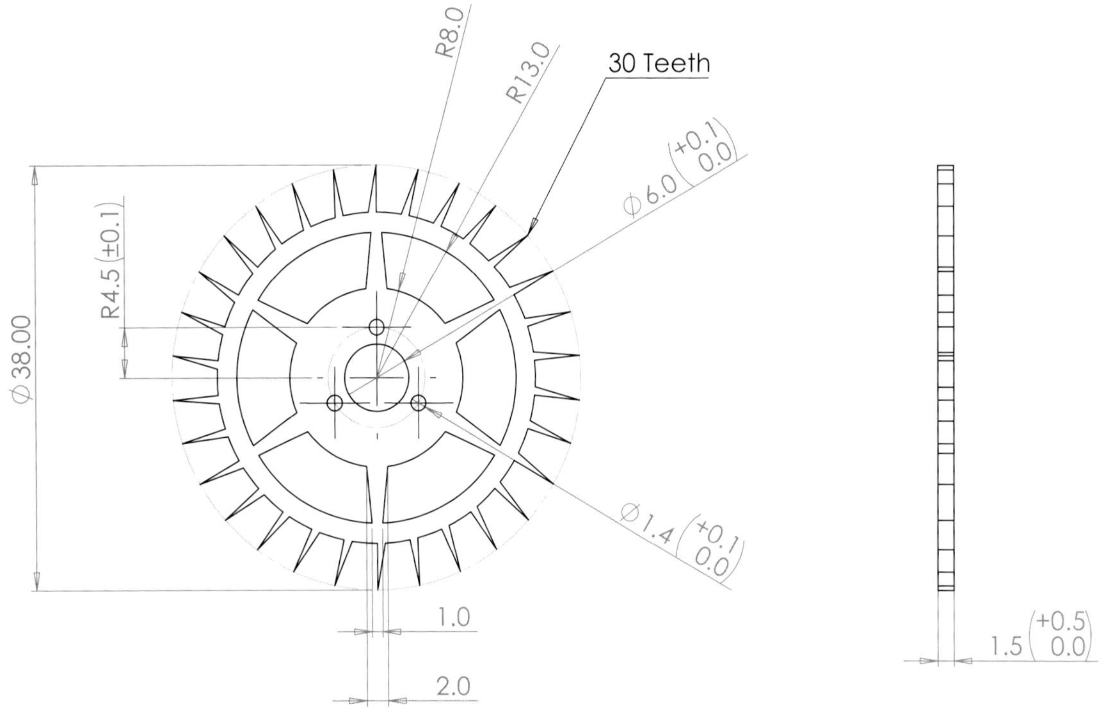

Deadbeat escape wheel.

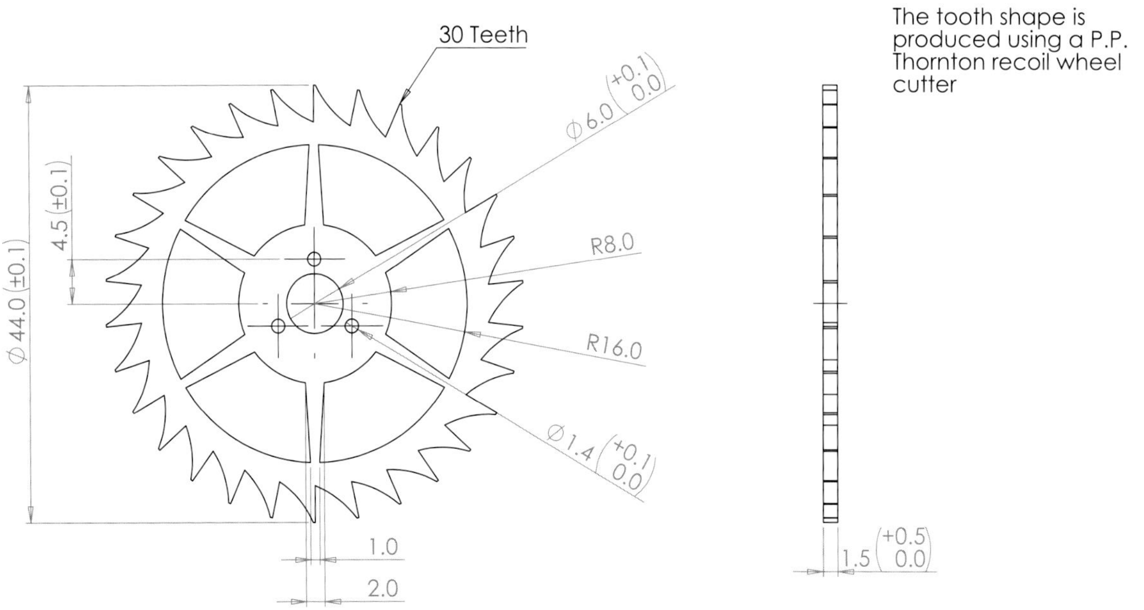

Recoil escape wheel 6 crossings.

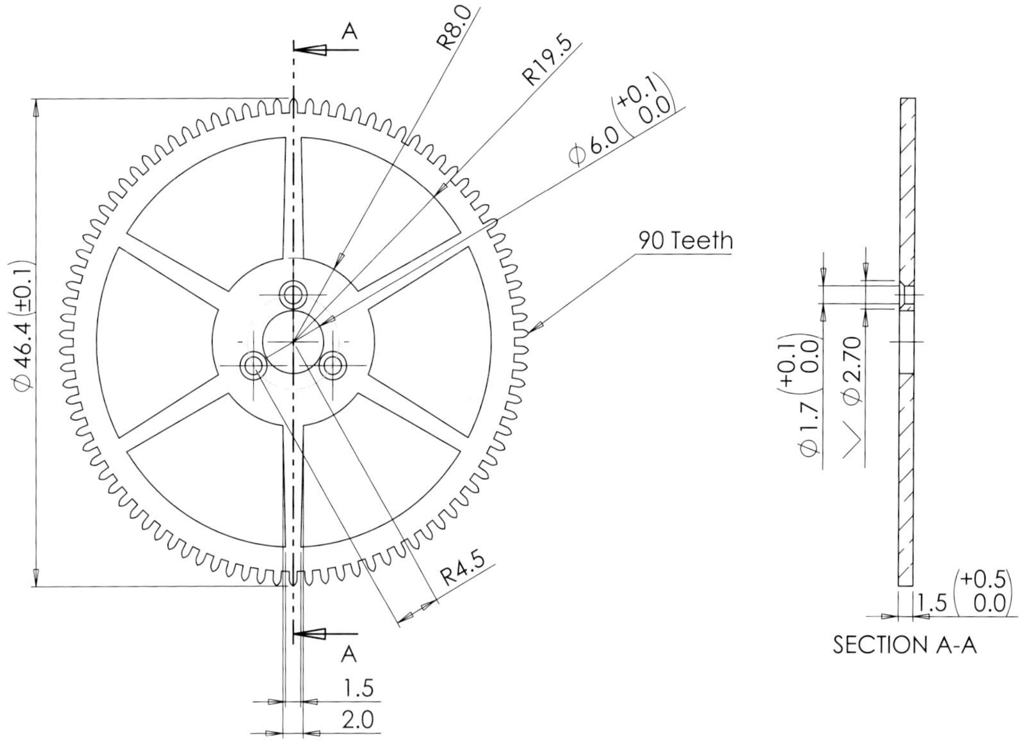

3rd wheel 6 crossings.

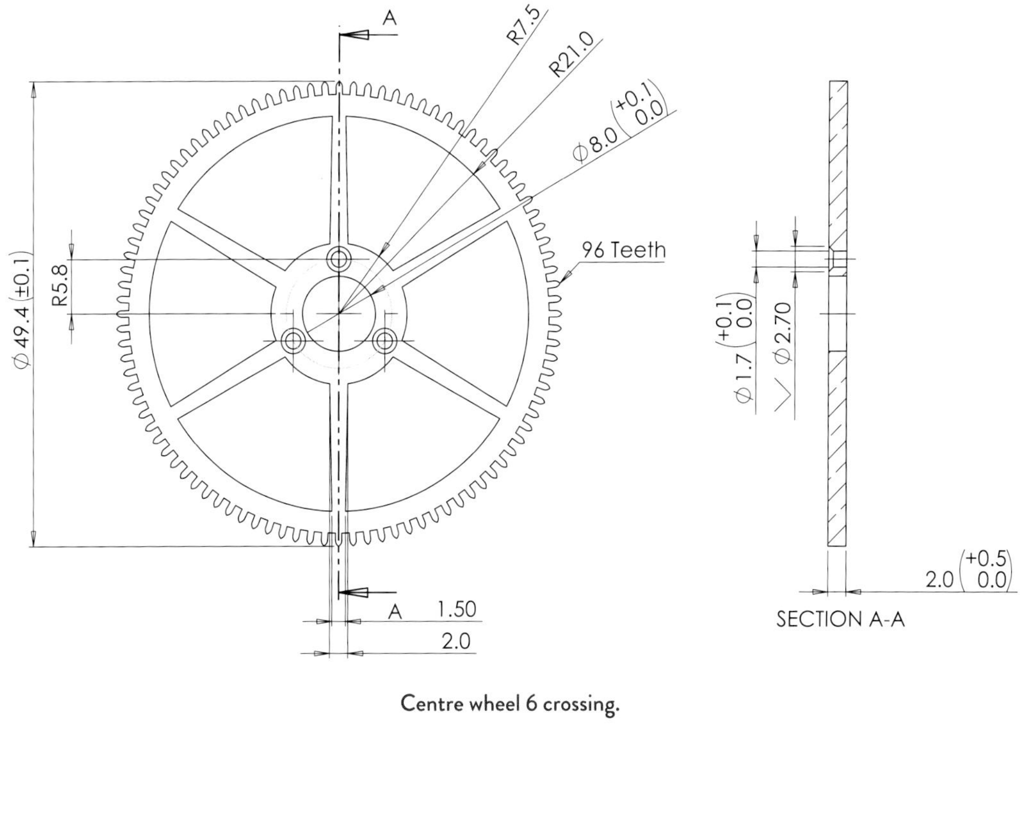

Centre wheel 6 crossing.

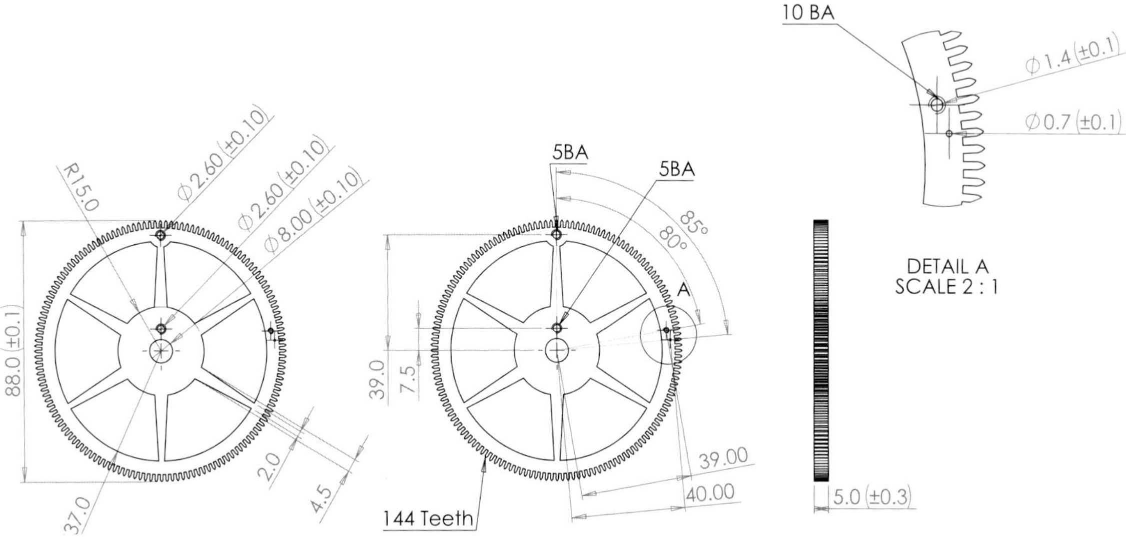

Great wheel 6 crossings.

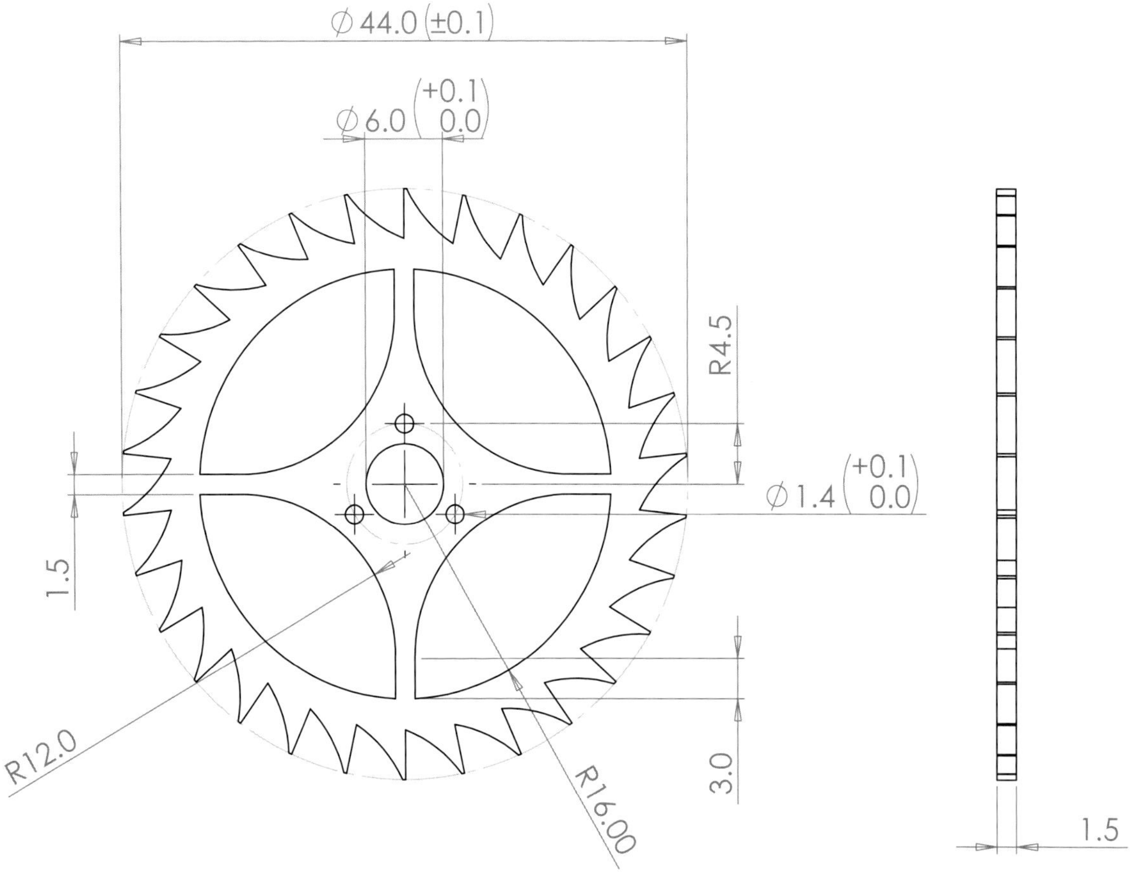

Recoil escape wheel 4 crossings.

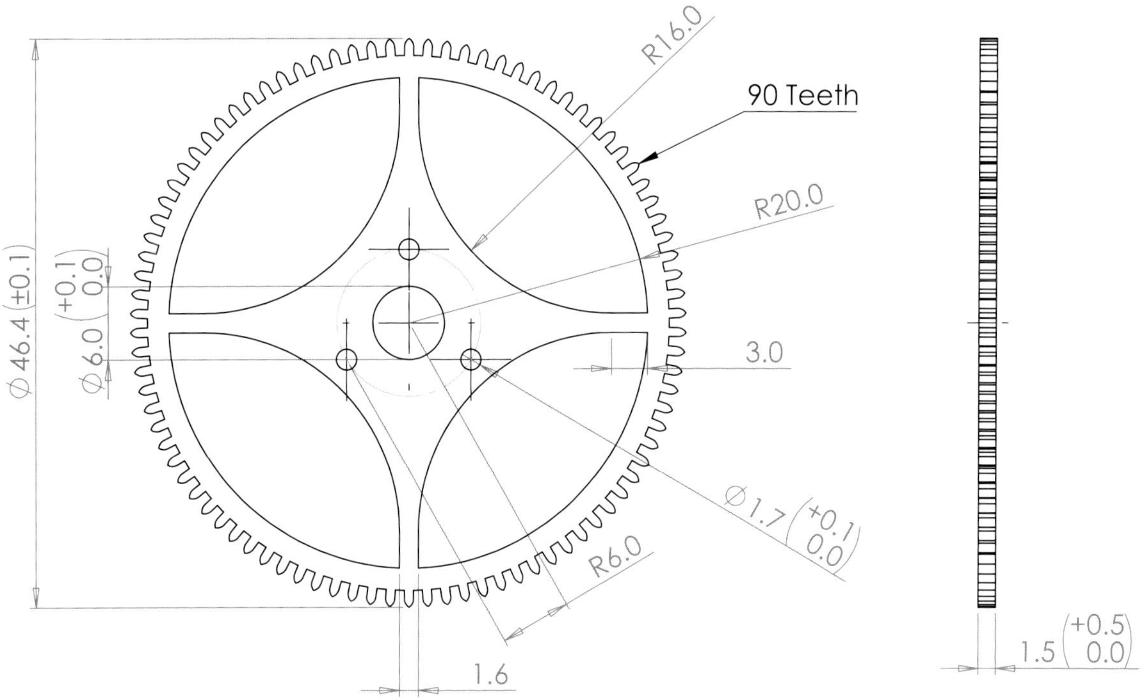

3rd wheel 4 crossings.

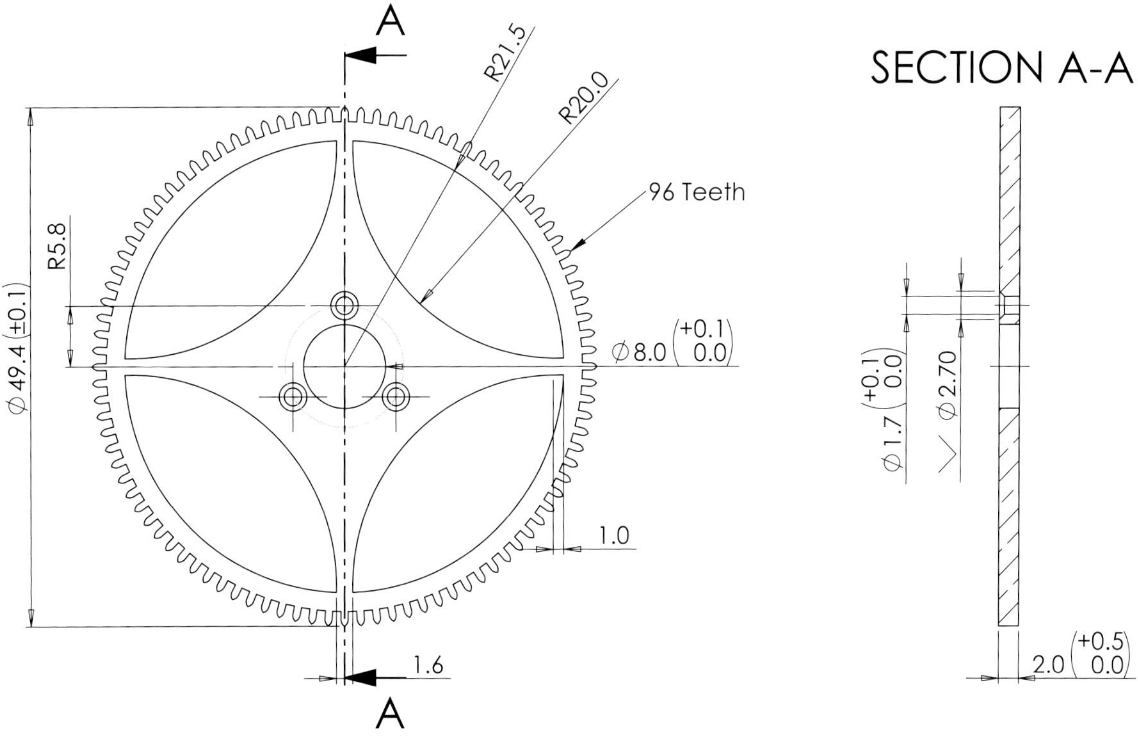

Centre wheel 4 crossings.

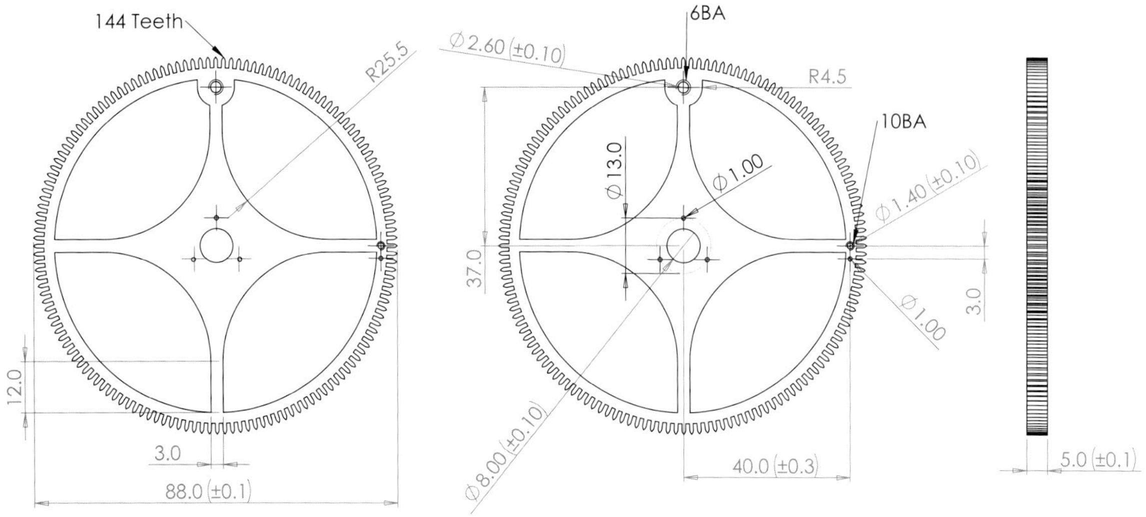

Great wheel 4 crossings.

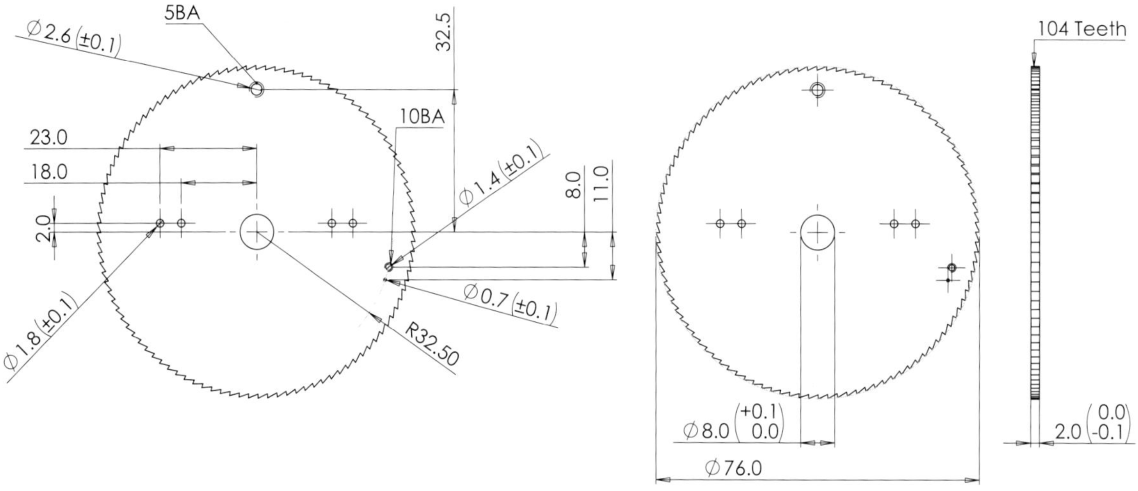

Maintaining power ratchet wheel.

It is set to leave just a small amount of thickness in the tip of the tooth to give it strength. The theoretical thickness of the tip of an escape wheel tooth is measured in degrees, and is 1 degree. So the actual tooth thickness for the recoil escapement for this project at 1 degree, is the diameter of the escape wheel – 44mm multiplied by 3.142 and divided by 360, which gives a tooth thickness of 0.38mm.

How cutters can look when correctly set. Here is the recoil escape wheel cutter viewed from above. This cutter has a sharp edge, which is readily lined up with the centre of the wheel blank.

This shows the deadbeat cutter, which by comparison has a flatter profile. As can be seen the left side of the deadbeat cutter – the vertical side – is lined up with the centre of the wheel blank to be cut.

For the deadbeat wheel it is 0.33mm, as the diameter of the escape wheel is 38mm.

Wheel Crossing – Marking-Out

Crossing a clock wheel is primarily done to lighten the wheel. Wheels can be crossed with a number of patterns, but the wheel should finish as light as is sensibly possible, whilst remaining strong and reasonably balanced. Traditionally clock wheels are 'crossed', as the spokes are called, with four crossings, although better-quality clocks tend to have five or six crossings used.

The lighter the 'train' of wheels the less force is required to drive them, and there is less wear on the point where the wheel and pinion engage. The going train in a clock is unique in that it never has the chance to run constantly. It only ever moves one tooth of the escape wheel at a time. Each audible 'tick' signifies the end of the movement. The train is then stationary, or recoils in the reverse direction, before being enabled to move just one tooth forward again. For each 'beat' of the clock the whole wheel train starts from a standing start, and stops again. It therefore stands to reason that the lighter and more smoothly the train runs, the better the clock performs, and less weight is required to drive it.

For this publication the movement fitted with the recoil escapement movement has the traditional four crossings, and the deadbeat movement is shown with six crossings. If you do not mind the extra work, six look elegant – although any inconsistency in marking out the six crossings will show.

Generally, the only wheels not crossed are the cannon pinion, reverse minute wheel and the hour wheel. That being said, these can be partly crossed to provide a counterweight to the clock hands.

If the whole movement will be visible when you finish your clock then the 'motion work', as this group of wheels is known, would look more elegant if crossed.

The first cut on the recoil escape wheel has removed a little more than half the brass necessary to form the tooth. As can be seen the teeth are thick and flat topped.

The teeth are cut to the correct depth; note the tip of the teeth are not sharp, but have a 'top' to them.

The deadbeat escape wheel has thinner teeth, which are therefore more fragile. Hence they are cut using three or four passes so that each cut is finer. Shown is the first pass.

The last cut forms the tip of the tooth. Aim to make this a shallow cut and use a smooth action.

Looking straight at the teeth of the deadbeat wheel whilst still mounted on the mandrel in the lathe.

Generally the cannon pinion wheel is *not* crossed, in order to give a good surface for the hand tension spring to run against. We will return to the hand tension spring in a later chapter.

To mark the crossings it is necessary to decide how the wheel will be mounted to the collet used to secure it to the arbor. In the majority of clocks, the wheel is riveted to the collet or pinion, but the author prefers to screw them in place. If you decide to cross the wheel with six crossings you might prefer to screw the wheels to the collet with three screws, in which case the screw holes need to be marked out as you mark-out the wheel blank. The location of the screws is shown on the drawing. It is quicker to cross and finish the wheel with four crossings, as there are just two curves to cut.

The chain-wound option for the great wheel shown with the traditional four crossings.

The great wheel crossed with six crossings.

If you have a depthing tool, this can be used to act as a divider to help mark out the wheels. If you are working without a depthing tool, use the lathe to make a brass centre to fit into the centre hole of the wheel. This is simply a disc of brass turned to the diameter of the centre of the wheel. Whilst the disc is in the lathe, use a small-diameter centre drill to mark the centre point. This can then be parted from the lathe the same thickness as the wheel and pushed into the centre hole of the wheel to give the central point. This acts as a centre point to locate the point of the divider, meaning a regular set of dividers can be used. Remember any marking lines will need to be removed, so the fewer and lighter they are made the better. The wheel can also be remounted onto the mandrel, or placed in a six-jaw chuck, and placed back into the lathe. Taking care not to mark with a deep line, the turning tool can be used to scribe across the wheel, and also to score a line round the wheel by turning the headstock by hand. With the wheel still in the lathe, and again using the tool very lightly, the wheel can be divided into four or six using the dividing head arrangement used to cut the teeth. If crossing with four crossings, divide the wheel into eight, as this will help with marking the inner curve in due course. Whilst marking-out in the lathe saves using dividers, care needs to be taken, as it can be difficult to make a light mark with the lathe tool.

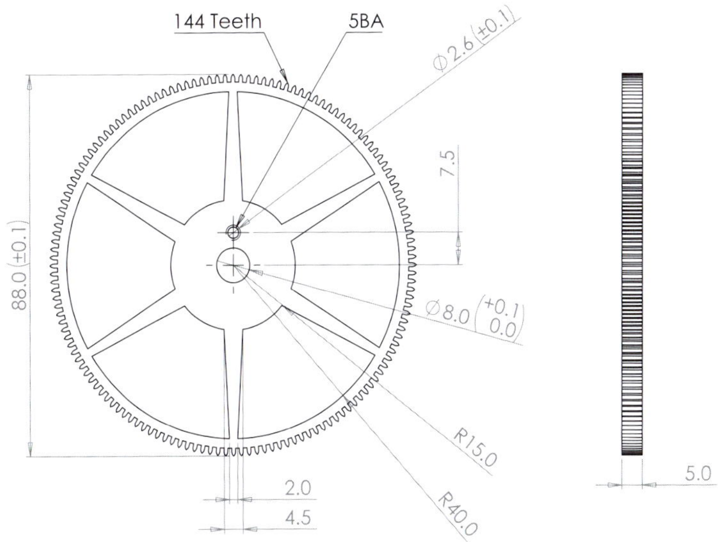

Great wheel for maintaining power.

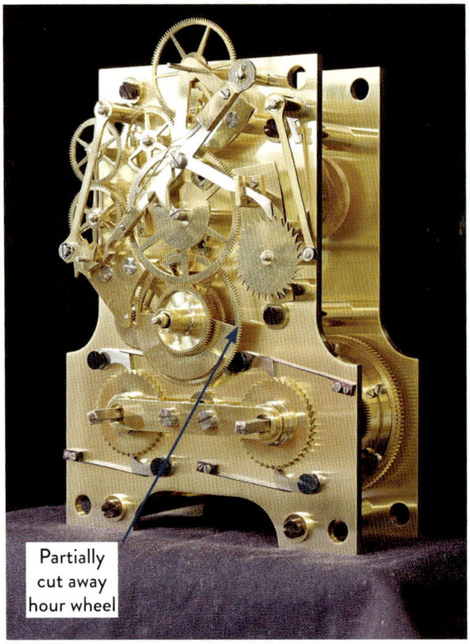

An example of the author's work showing partly crossed wheels. This is done to counteract the weight of the hour hand.

1

To mark-out six crossings, begin by marking the six initial points for the crossings. In this instance, they have been marked-out whilst the wheel is mounted on the lathe.

2

The next stage of marking the crossings either side of the initial centre lines. This is done once the wheel has been removed from the lathe.

When crossing with four traditional crossings, the inner curve from the inside of the rim to the centre of the wheel is not a true radius. You can mark it so, but this tends to leave quite a bit more brass and therefore weight in the wheel than is necessary. There seems not to be a rule for just what shape the inner curve should be. As long as it looks elegant and lightens the wheel – whilst remaining strong enough to carry out its required function, it will work. It is also good practice to make the wheel reasonably balanced.

For a suggestion, you might try the following for the great wheel: make a washer with a hole of diameter 8mm and outside diameter 15mm. Place the wheel on the mandrel and put the washer on top. Lightly scribe round the washer, to give an inner ring to join the crossing lines to. Looking at the marking out for the four crossings, join the opposite marks made for the four points going across the centre. You could say join 12 o'clock to 6 o'clock and 3 o'clock to 9 o'clock. Next, the in-between four marks need to be joined in the same way. So, thinking about the clock dial again, join the points where the hour hand would be at 1.30 to the point it would be at 7.30, and also where it would be at 4.30 with 10.30. These extra four points are used for marking out the inner curve with a ruler and scribe as the four crossings are made with two curves.

Mark 1mm either side of every other line so you have four evenly spaced crossings 2mm wide, and four evenly spaced marking out lines.

For the remaining four lines, place one of the divider tips where the crossing line meets the circle scribed for the curved section of the crossing, marked as point 'A' in the photo on the following page. Open up the divider so the remaining point touches the

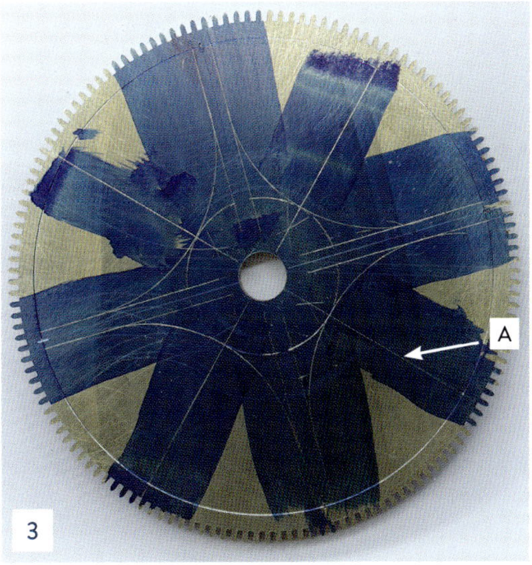

The great wheel with all the marking-out complete and ready to be cut out for the traditional four crossing pattern.

All six cuttings are cut and ready to be filed to shape.

15mm diameter scribed ring on the centre of the wheel. Scribe a line that will touch the 2mm marked lines. This will give an acceptable inner shape to the crossing (*see* photo 3 above).

It used to be reckoned it would take seven years to become a journeyman clockmaker. One of the many things you would learn along the way is how to cross wheels elegantly. It does take a bit of practice. This is where carrying out the initial suggested exercises with the piercing saw and files will pay off. If you are not pleased with your first attempt, you will have those spare wheels....

Wheel Crossing – Cutting the Crossings

Near the corner of each marked crossing centre punch a mark, and drill a 1.2mm-diameter hole. This hole is the starting point for crossing the wheels with the piercing saw. Insert the blade into each hole in turn, taking care not to cut into the marking out lines indicating the crossings, cut slowly and accurately. Time and care taken here will save a lot more time and possible disappointment with the file.

It is important to cut the curved section as smoothly as you can, as this greatly aids the shaping of the crossing with the file.

Having cut the crossings it is necessary to file them to shape before polishing them to make a pleasing finish.

As you push the file forward, work away from the corner so that the file finishes about halfway along the crossing as shown overleaf. Once the corners which can be filed in this way have been done – either four or six depending on the number you are crossing – turn the wheel round in the vice and file the remaining opposite corners. The curve being shaped is a smooth curve, so use light strokes and try to move the file smoothly as you work.

For the six crossings there are the flat tapered sections and the inner circle sections to file to shape. Place the wheel as low as you can in the vice with the section to be filed horizontal and file straight across. Take great care not to catch the file in the corners of the outer curved section when filing the flat sections of the crossings.

Filing and Polishing the Crossings

All the surfaces for the crossing are filed across to shape, and then draw-filed to smooth them down and to smooth the lines of the curves to shape. When finishing the crossings it is important to keep the corners nice and crisp if you can. One of the neatest ways is to use a graver – a lathe hand-turning tool – as a scraper to scrape the surface. This is another skill that requires care, but rewards patience. The object is to finally finish the crossing with a traditional pressure polish by work-hardening the final surface using a burnisher. Before we do this though, the surface is scraped with the edge of the graver. The author is sure there would have been a traditional tool for this, but has not managed to source one. A 'graver', which is used for hand turning on the lathe, will work well.

Hold one end of the graver in a pin vice, pass it through the crossing and, using its corner, work from the corner of the wheel crossing, draw it back towards the centre. Looking at the photo, the inner section facing which is catching the light shows how the surface looks when scraped. Do not apply too much pressure otherwise the surface may look rippled.

Now using the traditional burnisher, the wheel is placed onto something such as a piece of leather – so

Place the wheel in the vice, taking care to protect the wheel edge from vice marks with copper/brass jaws. Use a fine-cut, half-round file – shown is a Swiss number 4 cut. Place the edge of the file into the corner as shown.

Next, finish the sections between the corners with the same file. When this is completed the same file is used to 'draw' along the curves as shown. Take your time, it is better to do it slowly and once. Use the same file to shape the inner curved section of the crossing in the same way.

The inner section is shown being scraped with the graver. The crossing facing the camera, which is catching the light, shows how the surface looks when scraped.

Press the burnisher onto the surface of the brass and rub along the crossing to produce the finish shown.

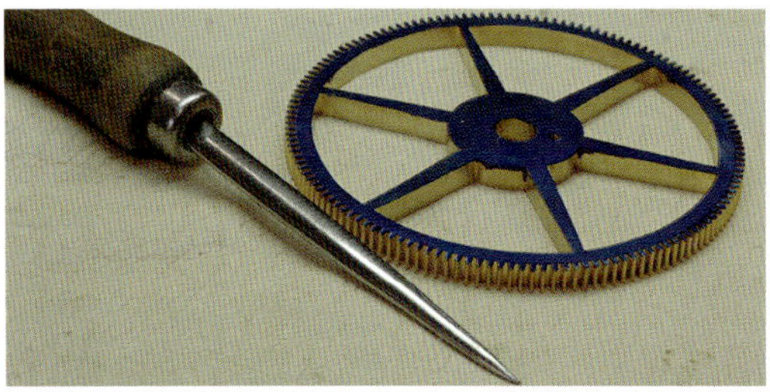

A traditional burnisher and the wheel ready to finish the wheel crossing laying on a piece of leather.

as not to damage the teeth – and the burnisher is pressed onto the surface of the brass and rubbed along the crossing. Take care not to slip off the rim with the burnisher, as this will mark the rim.

Work the corners and then the centre section of the curve, finishing with the straight sections on the six-crossed wheel. Each of the traditional four crossing sections has just the two curved sections to finish.

Having crossed the wheel the end is in sight, with just the sides to finish, using the micro-finishing gel sheets secured to a piece of plate glass. Take care to apply an even light pressure over the surface of the wheel, finishing the sides of the wheel to a flat polish.

Begin with the gel film marked 60 and use plenty of water to lubricate the film sheet. This helps to stop the

The corners and then the centre section of the curve are worked. The straight sections on the wheel are the final pieces to burnish.

Use the micro-finishing gel sheets and finish the sides of the wheel to a flat polish.

The finished great wheel.

The polished and mounted centre and third wheels. It takes practice and a bit of time, so to speak, but the finished wheels are worth the effort.

grain of the film sheet clogging, lubricates the cutting action and results in a more even cut more quickly.

Work through the grades of the film sheet until the desired finish is reached. The author finds it best to do several strokes, then rotate the wheel about a quarter of a turn, and then do several more. You may also find it useful to try moving the wheel round the film sheet. The main thing is to avoid digging the leading edge of the teeth into the film sheet, as this will give an unsightly chamfer to the tip of the teeth.

The remainder of the train wheels are made and finished as described above.

Do bear in mind the minute wheel, reverse minute wheel and hour wheel are traditionally not crossed in a typical weight-driven clock. If you are making the deadbeat escapement you could cross the hour wheel, either by crossing half of it to act as a counterweight for the hour hand, or simply cross it out. The author tends to cross the hour wheel, as it looks pleasing and saves a bit more weight. Do not cross the cannon pinion wheel, because a continuous rim is required for the hand tension washer to run on when the clock is set to time.

Pinions

7

PINIONS ARE READILY MANUFACTURED in the workshop, although it is more time-consuming than wheel cutting. It used to be possible to buy a product known as pinion wire. This was a length of steel with the profile of the pinion module required cut down its whole length. To produce the finished arbor and pinion, one would cut through the leaves of pinion either side of where the finished pinion head was to be located. The remaining leaves were then removed and the arbor turned to the required dimensions.

In today's world, horological steel pinion wire seems no longer available, so it is necessary to cut the pinions required.

There are two ways to produce the arbor with the pinion: one way is to cut the profile of the pinion directly onto the steel arbor blank. Due to the choice of pinion diameter for this project, a second method is also available (the loose pinion technique outlined on page 85), which is to produce a length of pinion from which a section can be cut, and a hole drilled down the centre to match the diameter of a plain steel arbor. The pinion is then secured to the arbor by using either a modern adhesive such as Loctite 638 or soft solder. It is easier to locate the pinion if a step is turned onto the arbor blank for the loose pinion to butt against.

If soft solder is used, care needs to be taken to ensure solder does not run into the pinion leaves. The flux will also need to be cured to prevent rusting.

There are various views as to the best steel to use, and whether to harden and temper the pinion and arbor when made. In the past the steels available were always hardened and tempered. The author's experience has shown that using EN8 or EN24 steel bar stock, as a basis for making the whole arbor, is a good compromise between the steel being reasonably workable

There are two ways to produce the arbor with the pinion. This method cuts the profile of the pinion directly onto the steel arbor blank.

This method produces a length of pinion from which a section can be cut.

The pinion is parted from the length of the pinion and a hole is drilled through the middle.

The general set-up for making the pinion blank. Note that the author's lathe used does not have a fitted lubrication feed; instead, a brush supplies lubrication for the cutter, and a steel tray collects the surplus lubrication with swarf to filter later.

but strong enough to function without wearing unduly when not hardened and tempered. However, EN8 and EN24 do harden and temper well when required. One drawback that can be avoided by not hardening and tempering finished arbors – especially the centre arbor – is the risk of warping, which then requires 'peening' (a hammering technique for straightening warped arbors) to correct the warping. In the interests of making the clock movement as straightforward as possible, and following much discussion with fellow professionals, the author has elected for this project that the arbors will not be hardened and tempered. In due course when the pallets are made these will be hardened and tempered and the technique explained.

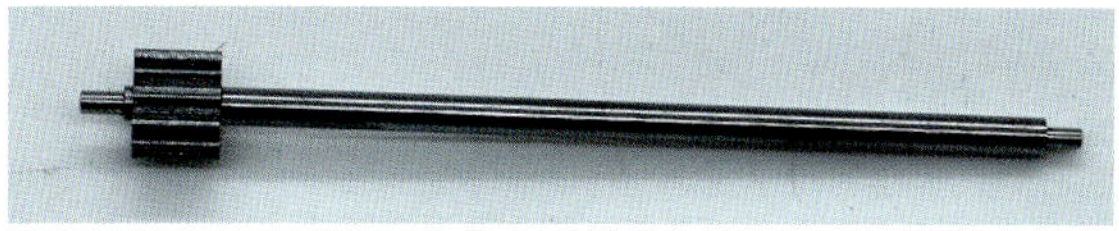

The pinion shown located on the arbor.

MAKING THE PINION BLANK

Note that unlike cutting wheel teeth in brass, lubrication is required to cut the steel for pinions and the cutter must be run much more slowly – at approximately 260rpm for a Thornton's cutter of diameter 24–26mm.

MAKING THE SOLID PINION AND ARBOR

The first task is to turn the arbor on which the pinion needs to be formed. It is best not to turn the whole arbor to begin with. The more steel remaining the stronger the arbor will be whilst cutting, and the better the finish will be when cutting the pinion. It is also necessary to support the end of the arbor that is not held in the headstock of the lathe. This requires making the following tailstock support.

This centre for supporting the pinion whilst it is cut is held in the tailstock to support the pinion blank.

THE TAILSTOCK SUPPORT

Make a centre for the tailstock as shown in the drawing overleaf. It is a turning exercise with a flat section filed onto the top of the support. A centre drill with an end diameter of approximately 1mm is used to drill the cone

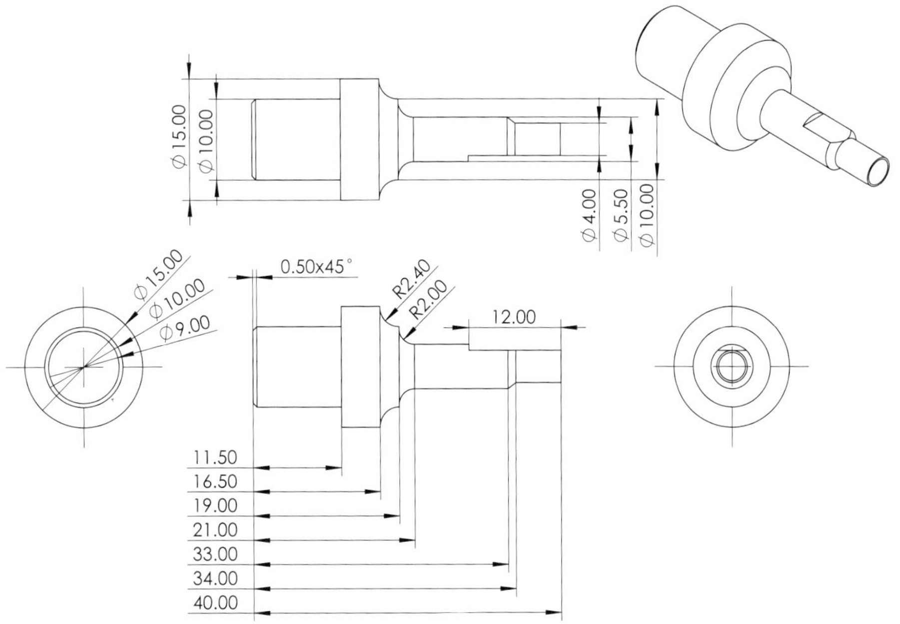

Tailstock arbor support.

shape in the end. The flat section is so the pinion cutter does not cut the support and enables you to see when the cutter is clear of the pinion leaf being cut, before rotating for the next cut. Whilst for smaller dimensions further supports would be required, for the clock being made here the tailstock support will be sufficient.

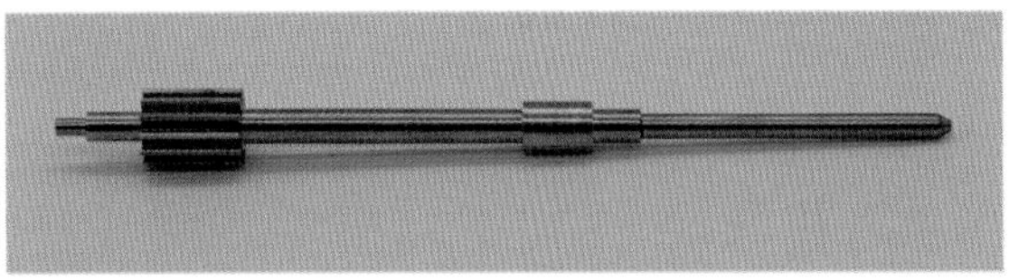

The centre arbor being produced, since this is the longest arbor to turn.

Returning to making the centre arbor, care is required as it will be necessary to change the position of the workpiece in the lathe. Begin by fitting a length of steel of diameter 10mm into the lathe with approximately 20mm exposed from the collet or chuck. Turn a point onto the end, this will locate the end of the arbor in the tailstock support previously mentioned. Try to match your point with the angle of the drill used to make the hole in the tailstock support.

Next turn the diameter down to give clearance for the bottom of the pinion leaf, which for the centre arbor will be diameter 3.8mm for the length of 10.5mm shown in the drawing. In due course the pivot will be turned on this end. Draw the steel through the collet so that around 50mm is now free from the headstock. In order

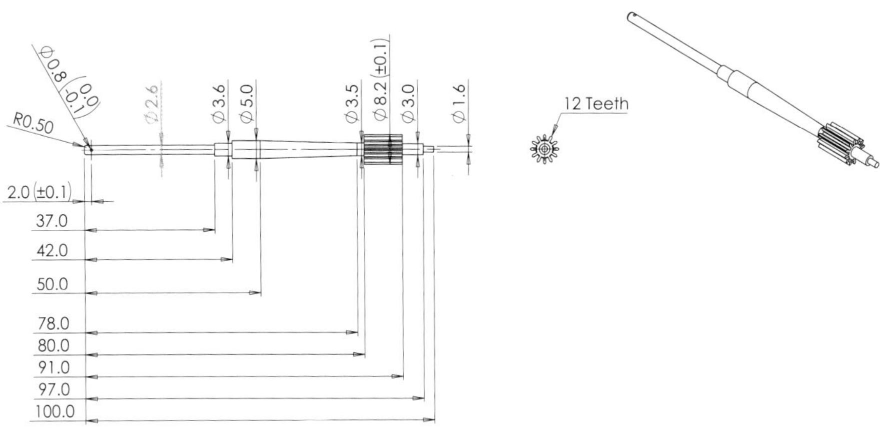

Centre arbor-solid pinion.

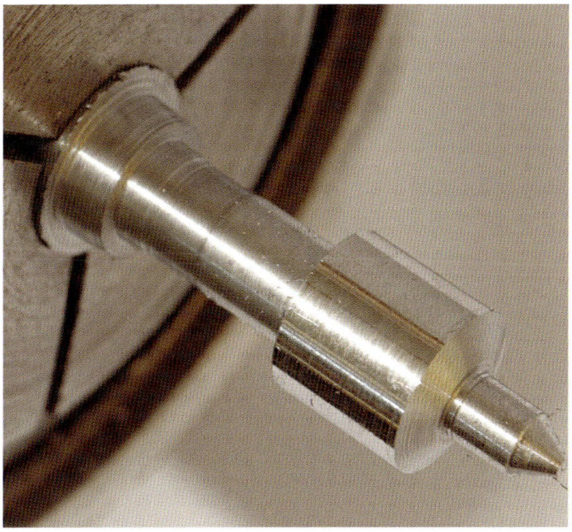

The arbor turned ready to cut the pinion. Note the end has a cone turned onto it to locate it in the tailstock centre shown. Beyond the section for the pinion itself to be cut, the diameter of the bar has been reduced only enough to give clearance for the cutter, to give as much support as possible during the cutting process.

to keep the workpiece as rigid as possible it is best to draw only as much steel through as needed for the cutter not to touch the lathe fittings when it passes through the pinion. Mount the turned point into the tailstock tool, making sure to lubricate it.

Now the diameter for the pinion itself is turned: 8.2mm for the centre arbor. Using a narrow tool in the lathe, turn beyond the length of the pinion to the diameter of 3.8mm for approximately 20mm. This allows the cutter to run through the pinion.

(If you wish, you could leave the arbor diameter at the pinion diameter and cut a longer length of pinion, in which case you would turn 26mm of length at a diameter of 8.2mm for the centre arbor. Although the arbor is stronger with the thicker diameter, this method does take longer and care will be required when turning the unwanted length of pinion away.)

Next, the pinion leaves are cut. The technique is the same as for cutting brass wheels, except that the cutter is run at a considerably slower speed of 260rpm and lubrication must be liberally used. In the author's workshop a steel tray placed under the lathe bed collects the cutting oil run-off and steel swarf. It can be filtered in due course and re-used. Take care to feed the cutter through the steel as smoothly as possible. As with cutting the wheels, the cutter must be central to the arbor on which the pinion is cut. This is important, as even a slightly offset cutter will produce pinion teeth that appear to lean with uneven off-centre sides.

To centre the cutter, withdraw the tailstock support and use the point previously turned on the end of the arbor to centre the cutter. Alternatively, remove the tailstock centre and place a dead centre or centring scope in the tailstock using the point to centre the cutter.

Once the cutter is centralised and locked in position the cutter depth will need to be set for the first cut. It will be necessary to make at least three cuts; the author used four in these photos to cut the pinion to the correct depth. The cut will be much neater using several shallow cuts and avoids damage to the workpiece or cutter by using too much force.

The first cut is a shallow cut, which does little more than indicate where the leaves will be. Cut all twelve leaves at this depth before lowering the cutter to make the second cut. As with cutting the wheels, it is not easy to say exactly how far to lower the cutter for each pass, but since several cuts will be made you should find it possible to gently come to the correct depth.

The pinion is the correct depth when the leaf has a tip. Make the last cut a shallow cut, to help give a better finish: it is an operation in which patience is rewarded.

The first cut is a shallow cut to indicate where the leaves will be. All twelve leaves are cut at this depth before lowering the cutter to make the second cut.

The arbor mounted and the cutter being moved into position ready to set the depth for the first cut.

The arbor diameter is turned to match the drawing.

The rear pivot is turned onto the rear of the arbor.

The end of the arbor is domed.

Having cut the pinion the arbor can now be turned. With the arbor mounted in the lathe, turn the section from the pinion to the mounting point of the centre wheel collet. This can be either a traditional taper as shown in the drawing on page 82, or parallel, as shown in the various photos.

The next section to turn is for the centre wheel collet to be mounted upon. The arbor is drawn through the collet to turn the next section. Continue to support the end in the tailstock. Care is needed, as the arbor is quite long for its diameter.

The front pivot is of sufficient diameter to enable the arbor to be large enough to mount the friction spring required to hold the hands in place. This allows the hands to be moved to set the time. In addition the front end of the centre arbor needs to have a large enough diameter to either drill a pin hole, or cut a thread onto it to hold the minute hand in place. For this clock we are using a front pivot diameter of 4mm. The centre wheel collet mount is therefore 5mm in diameter to give an adequate shoulder to the pivot.

The final section to be turned is from the front pivot to the end – what will be the front end – of the centre arbor. For this operation the arbor is once again drawn through

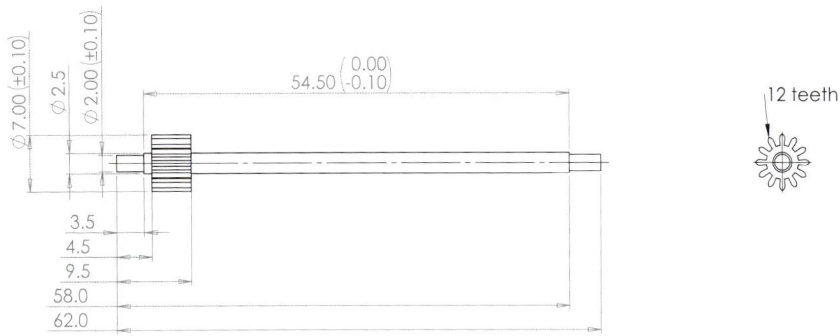

Escape and 3rd wheel arbor – solid.

the collet with sufficient length to turn the required length plus 2mm to allow for using the parting tool to cut the arbor from the bar stock. Take small – light – cuts with the tool so as not to apply undue force to the arbor.

Once parted off, the arbor is placed back in the lathe holding the pinion in a collet in order to turn the rear end section and the rear pivot.

One final operation whilst turning the rear pivot, and all remaining arbors and pivots, is to dome the end of the arbor. This can be done by either using a fine file or the lathe tool set at an angle. This allows for a neatly finished arbor in due course.

Turn the escape and third wheel arbors to the dimensions shown in the drawing on the previous page. Conveniently, they are both turned to the same dimensions.

USING THE LOOSE PINION TECHNIQUE FOR MAKING THE ARBOR

To save turning the whole arbor it is possible to make the pinion and add it to the arbor, provided the root (the centre) of the pinion has sufficient steel to allow for a hole to be drilled and the pinion to still have sufficient strength. This movement has been designed to give sufficient thickness to allow the use of loose pinions.

The procedure for this is to make the pinion as previously described. The difference is to cut a length of pinion long enough to produce two pinion 'heads' as they are called, one for the third and one for the escape wheel. Each pinion is then drilled with a hole to match the diameter of the arbor on which it is to be mounted. As can be seen from the photographs the author has turned a step onto each arbor to give a locating shoulder for the loose pinion to butt against. For the escape and third wheel arbors drill a hole 2mm in diameter. For the centre wheel pinion drill a hole diameter 3mm. These have been secured using an industrial adhesive, in this case Loctite 638.

The arbors were produced using steel known in the horological industry as Blue Steel. This is steel that is supplied hardened and tempered. It is used by horological repairers to manufacture parts for clocks and watches and it is supplied straight and true.

For dimensions to turn the various arbors, follow the drawings for the solid arbors, but use straight blue steel stock of the diameter of the finished arbor. For the escape and third wheel arbors use blue steel of 2.5mm diameter. Turn the pivots as shown in the drawing, and then on one end of each a step of 2mm diameter a further 6mm along the arbor from the shoulder of the pivot as shown in the top-right photo.

The three stages of making an arbor using a loose pinion:

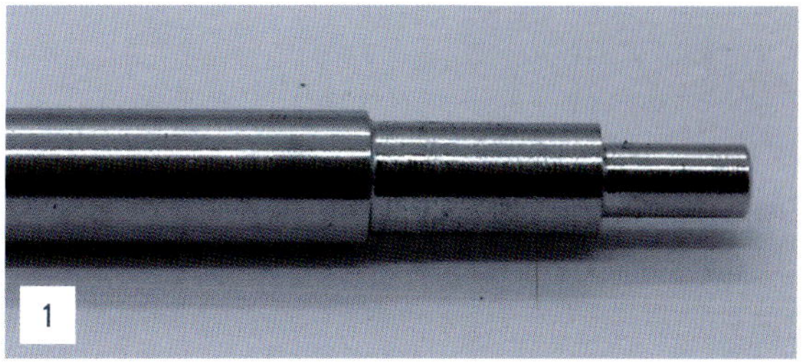

The steel arbor with a step turned of a diameter to match the hole drilled through the pinion head.

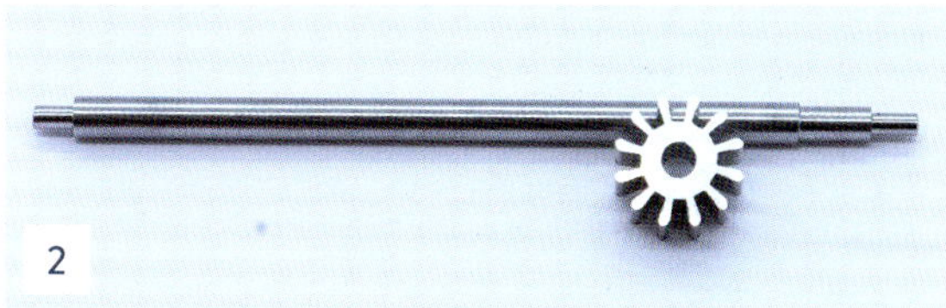

The arbor finished and the sides of the loose pinion polished ready for assembly.

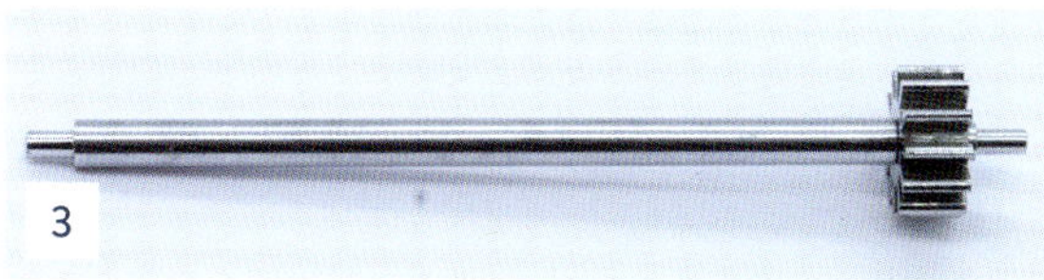

The pinion fitted to the arbor. Here it is secured using Loctite 638 adhesive.

For the centre arbor use diameter 3.5mm, turn the pivots as the drawings show, and turn a step a further 17mm along from the shoulder of the pivot to diameter 3mm.

Having drilled the pinion heads and turned the steps to accommodate them on the arbors, the pinions will slide onto the arbor and stop at the step. Finally, test the fit and secure the pinions with either Loctite 638 or soft solder.

The centre arbor is produced by fitting a loose pinion at one end and a steel collar in the middle to act as the end shake collar. The collar consists of a length of steel 4mm long with an outside diameter of 7mm.

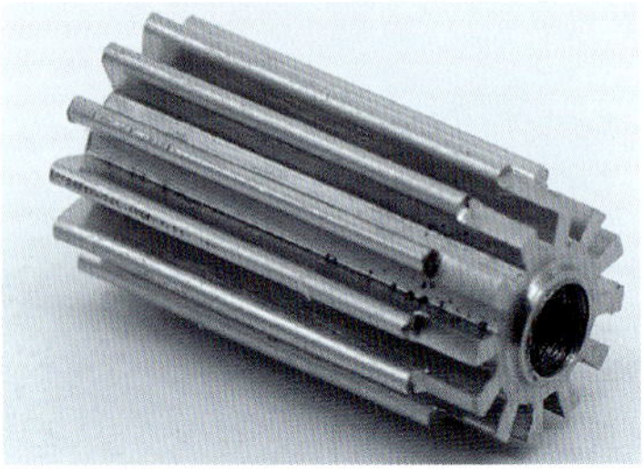

The reverse minute wheel pinion, shown with the leaves polished and the end diameter reduced ready to fit the reverse minute wheel.

The reverse minute wheel being lined up ready to be pushed onto the pinion.

The finished reverse minute wheel and pinion assembly.

Drill a hole through the middle of diameter 3.5mm to fit the blue steel used for the arbor. Remove the blue colour using a fine grade of gel sheet stuck to an old file – a buff stick – before securing the collar in place with Loctite 638 or similar.

Note: One point to note is that the hardness of the steel makes a tungsten carbide drill bit a useful tool for drilling the hole for the hand-securing pin. These are readily available.

One further pinion is required for the movement, the reverse minute wheel pinion.

Follow the dimensions shown, cutting the pinion as a longer section to give the extra length required for the pinion to be the mounting for the reverse minute wheel.

FINISHING THE ARBOR

To finish the arbor, use the self-adhesive abrasive film stuck onto hardwood lengths or onto an old fine file. Begin with the 40h then 30h, etc., finishing up with the pink-coloured, very fine film sheet. If using wet and dry papers begin around the 400-grade mark and work up from this point using 600, 800, 1,200 and 2,000. Having reached the 2,000 grade a final finish can be added using a piece of leather secured to a piece of wood, with cream metal polish such as Autosol. If polishing in the lathe do not forget to cover the bed of the lathe to protect it.

The pivots are best polished in the traditional way, using a wooden block, a pin vice, and a hand pivot burnisher/file. This is the reason for the exercise of filing pins at the beginning of the book. If you skipped the filing exercise, the author suggests you go back and do it rather than risk spoiling your hard work getting the pinions and arbors this far.

Place one end of the arbor in the nearest-fitting pin vice and tighten it in place. Using the file end of the burnisher and, lubricating it with white spirit or a light lubricating oil, place the tip of the file onto the pivot and, whilst applying a light pressure, push the file away from you whilst turning the arbor towards you. Do not push the file without rotating the arbor, as this will cause a flat to be filed. The object is to remove any marks remaining from turning the pivot. Use a light pressure on the file, as this will give a better finish and avoid the file clogging.

Having removed any turning marks, use the same technique with the burnisher end of the pivot file, again lubricated with white spirit, to polish the pivot. This helps to work-harden the surface of the steel as it is polished.

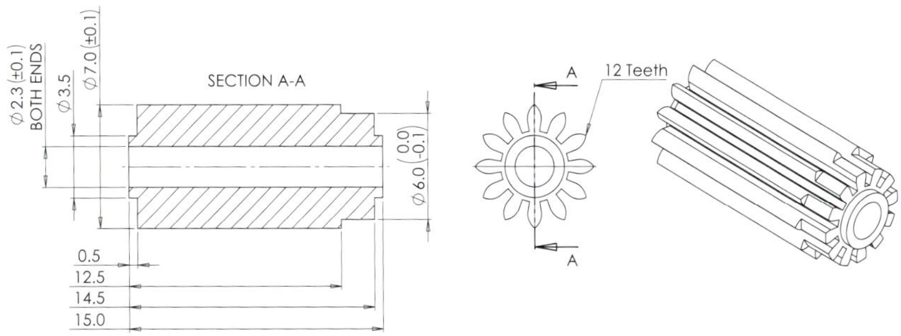

Reverse minute wheel pinion.

8 Assembling the Wheels and Pinions

HAVING MADE THE WHEELS and pinions, they need to be assembled by mounting the wheels onto the arbor with the correct pinion and then setting the completed arbors into the plates. Each wheel has to be mounted either onto a brass mounting known as a collet, or riveted onto the back of the pinion cut onto the arbor.

The centre wheel screwed to the collet mounted on the centre arbor, which is complete with its pinion.

The collet mounted onto the arbor.

12BA screws have been used to mount the wheels of the clock. The wheels can either be screwed or riveted to the collet, but need to be riveted to the pinions. In this project all the wheels, apart from the reverse minute wheel, are mounted on brass collets rather than riveted to the pinion.

Whether the wheel is riveted or screwed to the collet, the collet is mounted onto the arbor, and then turned true before mounting the crossed and polished wheel. This is to make sure the seat to mount the wheel is running as true as possible.

MAKING THE COLLETS

The procedure for making all the required collets is the same. Begin by making the third and escape wheel collets as they are the same dimensions.

Mount a piece of 12mm-diameter brass bar in the lathe, either in a collet or three-jaw chuck. (If you have suitable collets, use them in preference to the chuck.) Turn the brass to match the dimensions shown. When turning brass we tend to use around 1,100rpm on the lathe – this gives a good finish. Next, the centre hole is drilled: slow the lathe down and using a centre drill in the tailstock, with a tip diameter of 2.4mm (this is a standard centre drill found in the Dormer five-piece set), make an initial hole. Next use a drill to match the diameter of your third wheel arbor – in the drawing shown as 3mm – and drill a hole approximately 12mm deep into the brass. Now change the turning tool for the parting tool and part the section off 11mm long. You will now have one roughly cut-to-size brass collet blank with a hole matching your arbor through the middle.

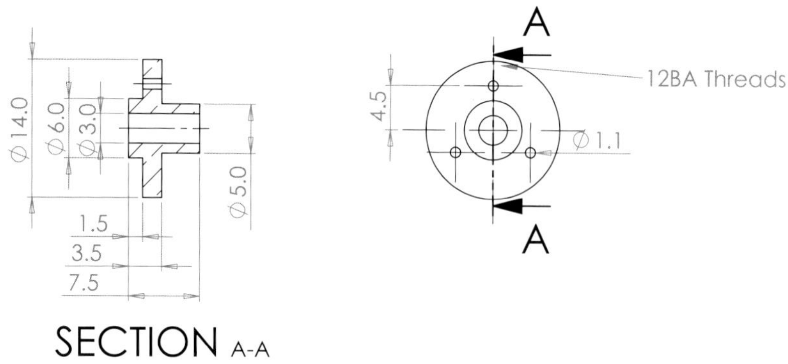

3rd and escape wheel collet using 12BA screws.

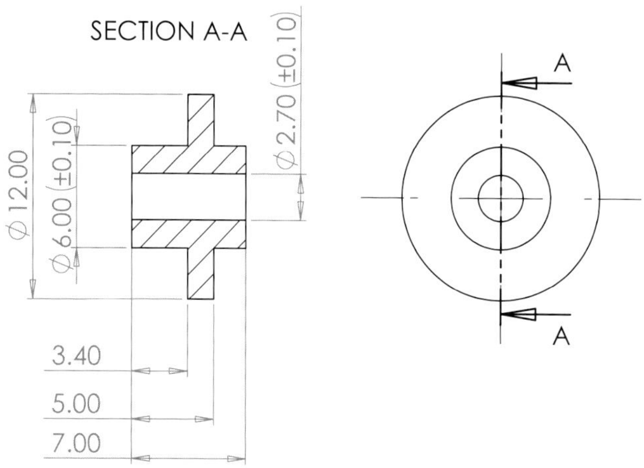

3rd and escape wheel collet – riveting.

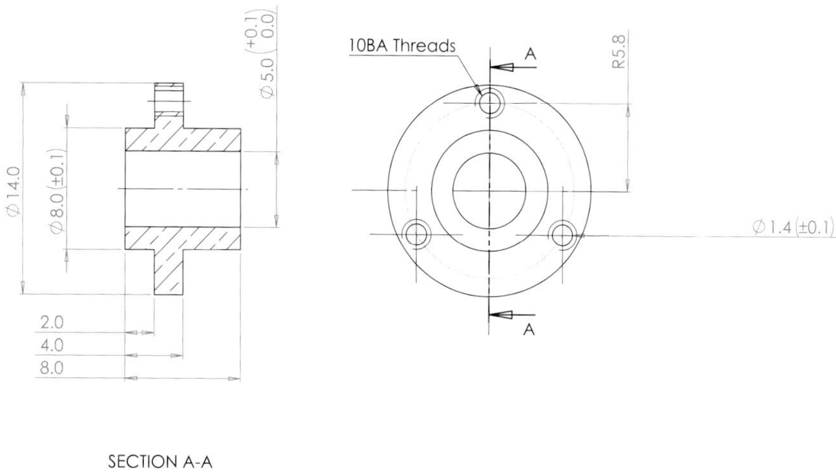

Centre wheel collet.

The centre wheel collet turned, with a hole drilled through, the diameter of which matches the centre arbor. The collet is ready to be parted from the bar stock at this stage.

MOUNT THE COLLET ONTO THE ARBOR

Traditionally the collet is secured to the arbor using soft solder. There are also several modern adhesives that can be used; Loctite 638 is used in this publication. The procedure for soft soldering is described next.

Soldering is generally described as 'establishing a firmly bonded connection by using solder material, where only the solder reaches a molten condition'. With temperatures up to 450°C, this process is termed soft soldering; with temperatures in excess of 450°C the proper term is 'brazing'. When the author was an apprentice, soft solder as a material had a high lead content and was often referred to as lead solder.

The collet is placed in position on the arbor and being heated ready to apply the solder.

The solder being applied to the collet. (Note the burner is not burning in the photo.)

These days the material used varies depending on what is being joined. Some examples of soft-solder are: tin-lead for general purposes; tin-zinc for joining aluminium; lead-silver for strength at higher than room temperature; cadmium-silver for strength at high temperatures; zinc-aluminium for aluminium and corrosion resistance; and tin-silver and tin-bismuth for electronics.

To soft solder the collet in place, slide it onto the arbor and apply a flux to the area to be joined. The job of the flux is to clean the surfaces of the two metals being heated to be soldered, and to break down the surface tension in the solder when heated to a liquid. It also helps prevent fresh oxides forming, so you should end up with a clean and secure joint. Put the collet blank in position on the arbor and heat sufficiently to enable the solder to melt when it is touched on the point where the joint is to be made. Do not overheat the work, and do not heat the solder directly, as it will melt before the collet and arbor are hot enough. This would result in the solder not flowing completely through the joint and then failing when the movement is running. It is the heat from the brass and steel that melts the solder, not the flame from the heat source. Illustrated is using a methylated spirit burner, as this will supply enough local heat without the ferocity of a blowtorch. Do not use an electrician's plug-in soldering iron: it will not work, as the metal is insufficiently heated.

Having soldered the collet onto the arbor, it will be necessary to remove any surplus flux and clean the joint. As the flux used by the author is water-based, it can be washed off. Some types of flux require neutralising using an alkaline such as ammonia. If doing this plenty of ventilation is required and the alkaline must be disposed of correctly. Make sure the workpiece is thoroughly dry.

Return the arbor to the lathe to finally turn the collet to the finished diameter shown in the drawing. Leaving the final turning until the collet is mounted ensures the seat for the wheel will be true. It would be very difficult, if not impossible, to depth a wheel to run (mesh) successfully against a twelve-leaf pinion if it were mounted at all off-centre.

Although the wheels can be mounted with screws, or by riveting, the two main advantages of screwing the wheel into place are the ease with which it can be removed if required, and it avoids the risk of damage when riveting into place.

Screw the wheel to the collet

To screw the wheel onto the collet the procedure is as follows.

The wheel has already been marked with the points to drill. Centre-punch these three holes and drill one with a 1.0mm diameter drill to use as a marker hole for the collet. Place the wheel on to the collet and carefully drill through the hole to mark the collet.

To illustrate this, the arbor – in this case the pallet arbor – is held in a split brass stake, the construction of which is described shortly. The crutch is placed over the collet and a drill passed through the hole in the crutch in order to mark the correct position to drill through the collet. The hole is then drilled with the correct diameter drill to allow it to be threaded using the correct tap. (The construction of the crutch is dealt with in due course.)

The collet for mounting the traditional crutch to the collet. It is firmly held in the split stake, in the machine vice ready to drill the pilot hole to locate the first screw.

Having used the drill to mark the location of the first screw hole, remove the wheel and drill right through the collet and, using a 12BA tap, thread the hole. Next, open the diameter of the hole used in the wheel to mark the location to 1.3mm diameter to allow the 12BA screw to pass through. Screw the wheel onto the collet. The wheel is now screwed securely to the collet with one 12BA screw. Using the 1mm diameter drill, drill the remaining two holes right through the wheel and collet. Remove the wheel, tap both the holes in the collet with the 12BA tap, and open the diameter of the two remaining holes in the wheel to 1.3mm for the 12BA screws to clear. Countersunk screws look neatest for mounting the wheel, so countersink the three screw holes on the wheel using a small countersink. Secure the wheel with the three screws, which should pass neatly through and secure the wheel to the collet. Remove any extra length of screw thread that may show beyond the back of the collet, holding the screw in the lathe to aid cutting any extra length away.

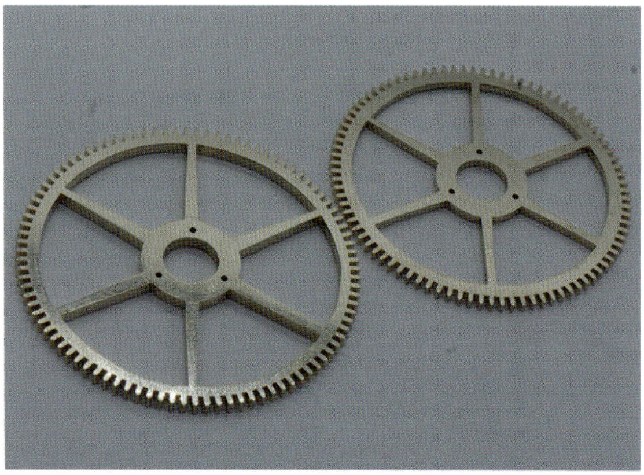

The centre and third wheels shown drilled with 1mm-diameter holes ready to line up on their respective collets.

Any extra length of screw thread that may show beyond the back of the collet is removed, holding the screw in the lathe to aid cutting any extra length away.

Riveting the Wheel to the Collet

If you have chosen to rivet the wheels to the collet then a split stake is required to rest the underside of the collet on, in order to rivet the wheel into place. A split brass stake is a simple tool that can be made in the workshop.

Make a Split-Staking Tool

The split stake is placed in the vice. Use the stake held fully in the jaws.

If you have a milling machine, this tool works best with a step milled down each side, as shown. If you do not have access to a mill, simply proceed without including the step: it will work well enough to be really useful.

Take a section of square or rectangular brass bar of approximately 12mm square and 90mm long. The actual size is not critical. Place it in the mill and mill a step down one side; again the size of the step is not vital – that shown is 3mm. Remove from the mill and cut the length in half, giving two sections 45mm long. Tidy the ends of the cut bar. The author used a collet in the lathe to 'face' the ends.

To make the two steel guide pins, take **one** brass section and drill two holes right through it – as shown

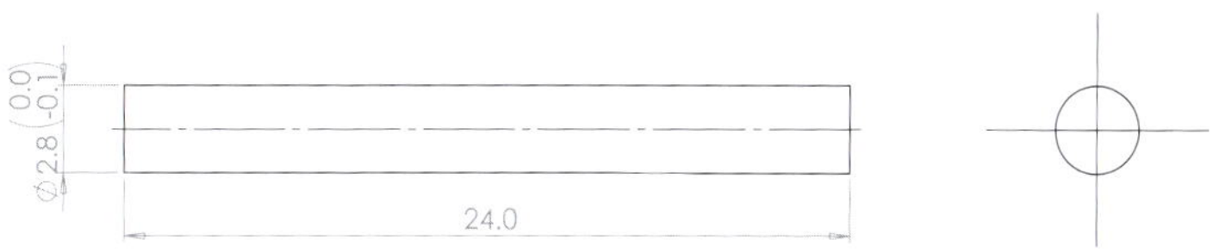

Split stake pin.

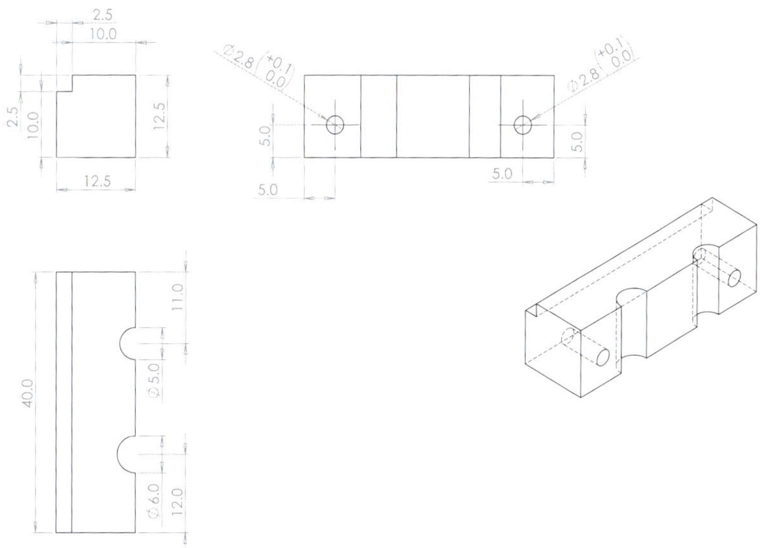

Split stake left side.

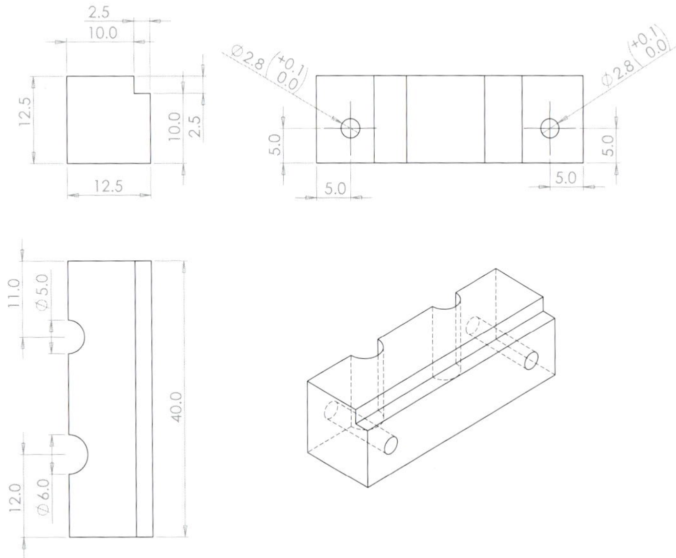

Split stake right side.

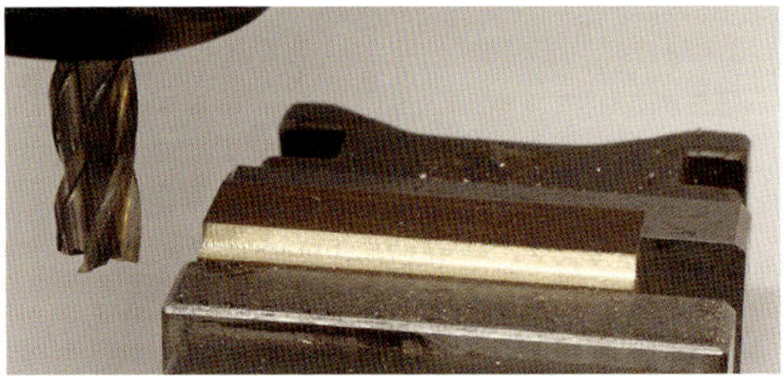

Milling a step onto the underside of the stake.

The marked brass section with the centre punched ready for drilling.

Place each section of brass in the lathe and face the ends to make a tidy job as shown. Note the two steel guide pins; these have their ends domed to make the tool more pleasant to use.

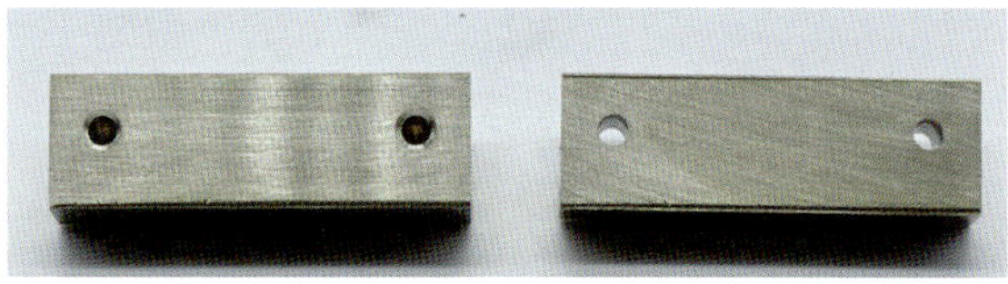

Both parts of the stake. It can be seen that the holes on the left-hand section are blind – that is, they do not go all the way through.

in the drawing – of 3mm diameter or to match a small diameter rod you may have in the workshop.

The diameter of the rod is not critical, but the fit into the holes is important. In the photograph blue pivot steel has been used. Drill through both these holes and then place both sections in the vice so the stepped side is towards the vice jaws, and therefore the unstepped sides meet in the middle and are clamped together as you close the vice. On closing the vice, it will become apparent that the milled step holds the sections at the same height and stops them sliding through the vice when it is used. Using the drill as a marking tool in a pin vice, mark through the two drilled holes to locate the centre point of the holes on the opposite section of brass. Remove from the vice, centre-punch and drill these two holes, stopping just short of passing right through. It is not the end of the world if you drill right through.

The depth to drill to is not critical – just try not to drill right through. Set the depth stop on the pillar drill to ensure the hole does not pass right through. Drill both holes to the same depth.

Next, cut two sections of the small diameter steel bar selected; these will pass snugly through the holes drilled and line up both sides of the 'stake'. Secure one end of each steel pin into the blind holes, using Loctite 638 or similar for this. Ensure the Loctite has cured completely before assembling the tool.

Place the stake back into the vice and gently file across the two halves to make sure the surfaces are true and line up. The finish can then be made smooth by draw-filing and finishing by running the surfaces over the gel papers attached to the glass plate. Finish to a level you are happy with.

Now consider the diameter of the various collet tail sections to be made. The object is to produce a two-piece stake that will fit snugly around the collet tail, giving a good surface for riveting the wheel into place.

The finished split stake.

Finishing the stake using gel sheets secured to plate glass, and used lubricated with water.

The stake separated so the hole can be seen.

The underside of the finished stake.

Assemble the stake and clamp it in the vice. Centre-punch in 15mm from one end to drill a hole that will be centred on the joint between the two sections of brass. Drill a hole through 0.2mm larger in diameter than the tail of your collet. The photo at the top of this page shows this.

Remove the assembled stake from the vice, tidy and finally finish the top surface, taking care to remove any burring that may have occurred. This is the surface that the collet will sit against when riveting. (Any marks left on this surface will be reproduced.) Use the gel sheets – wet – and wash off any residue thoroughly. Dry really well to avoid the pins rusting; do not forget to dry the holes as well!

You will now have a two-piece brass stake that will close with both sides lining up correctly. It makes riveting all the wheels in place a straightforward operation.

Back to riveting the wheel in place now we have the tool for the job

When turning the collet to the correct size, it is important to ensure that the diameter of the wheel mounting point is turned to the correct diameter in order for the wheel to be a snug fit. If the wheel is a poor fit, make another collet. It never works trying to rivet a wheel into place when it is not a good fit to the collet seat. (The result is a wheel that generally does not sit flat, and will probably be set off-centre.) Take a steel punch with a flat, smooth acting surface and gently tap once at a point on the collet rivet. Make another gentle punch 180 degrees round, and then at each of the 90-degree points. You should now have four light punch marks gently gripping the wheel into place. After the initial four light punches, if the wheel is not sitting flat it may be that there was either a burr on the edge of the wheel centre hole, or that the collet seat had not been turned right into the corner. In either case, the idea of the initial punch blows being light is to make it possible to place the assembly in the lathe, and turn off the light rivets if necessary. This releases the wheel from the collet, and gives a second chance to fit the wheel before the work is spoilt. (It is always disheartening to have to re-make parts because something went wrong.)

Provided the wheel is sitting flat and completely against the shoulder of the wheel, continue with the punch, working round the collet until there is a continuous neat ring of smooth punch marks and the wheel is firmly gripped.

Finally, to neaten the fitting, the mounted wheel can be returned to the lathe and the riveted surface turned to remove the rivet marks. Take care not to turn too much away, as it will result in the wheel not being held securely.

Place the arbor and collet into the stake.

Work around the collet with the punch.

Once the collet has been punched the whole way round to rivet the wheel firmly and flat onto the collet as shown, return it to the lathe and turn to remove the rivet marks.

The collet riveted surface following turning to finish. (It is shown held in the split stake for convenience.)

9 Making the Barrel Assembly

WEIGHT-DRIVEN CLOCKS ARE wound by lifting a weight, either by turning a key or by pulling a chain or cord – *see* photos. In the world of antique clocks, those of eight-day duration or longer tended to be key-wound, and those of 30-hour duration wound by pulling on a chain or rope to lift the weight. During the late nineteenth century, eight-day, mass-produced weight-driven clocks began to be chain-wound. This method is cheaper and quicker to construct, but perhaps not as satisfying to use. The main difference in construction between a clock wound by pulling a chain rather than winding with a key, is that the great wheel is mounted solidly on the arbor in the case of the chain/rope-wound clock. With the key-wound barrel, the barrel is solidly mounted on the arbor, with the great wheel being able to rotate – or rather the arbor being able to rotate – in the centre of the great wheel when the clock is wound.

The finished chain-wound barrel assembly.

The complete key-wound barrel assembly including the maintaining power variation.

Both barrel types are described in this chapter. If making the deadbeat escapement you will need to follow the extra work for making a maintaining power device – in this instance Harrison's design. Maintaining power is shown for the key-wound barrel in this project.

BEGINNING WITH THE KEY-WOUND BARREL

The general layout of the barrel to be constructed is shown in the photographs. Follow the correct drawing for the arbor you are making. There is the option for the provision of a maintaining power device for the key-wound barrel. If making the maintaining power, follow the arbor drawing for this option.

Begin by turning the barrel arbor to the dimensions shown.

Place a bar of EN8 steel of diameter 12mm in the lathe and turn the section from which the winding square and front pivot will be formed to the measurements in the drawing.

Achieving a better finish whilst turning the arbor will result in less time needed to finish the bigger diameter pivots, so use a higher speed lubrication and take small

The key-wound barrel with the extra maintaining power ratchet wheel fitted.

The rear view of the same barrel without the maintaining power ratchet wheel fitted.

cuts. Also turn the slightly larger diameter – which will enable the mounting of the barrel and locate the front pivot end shake shoulder.

The next procedure is to part the bar at the total length of the barrel arbor using the parting tool.

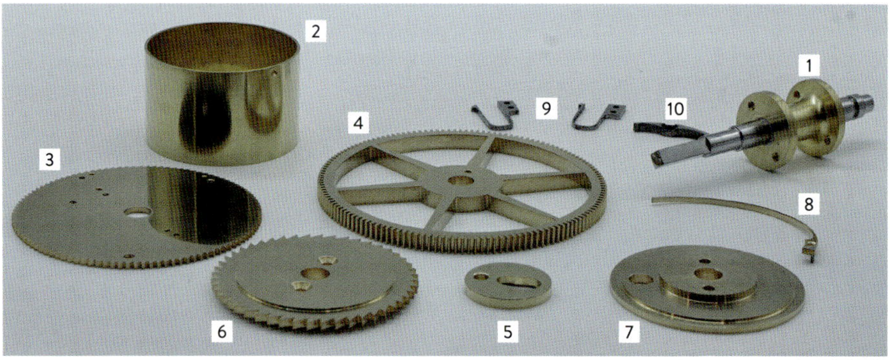

The parts required to make the key-wound barrel including the maintaining power parts. The barrel consists of the following parts shown numbered in the photo above:

1. The barrel arbor: this is the central arbor running through the whole barrel assembly. It has the two pivots turned onto it that pass though the plates; these form the running surfaces for the bearings. Finally, on the key-wound arbor, the winding square – in this instance to fit a number 12 key.
2. The barrel around which the gut or bronze line is wound. (Not steel line.)
3. The maintaining power wheel, if being fitted.
4. The great wheel.
5. The locking circlip. The great wheel and the locking circlip are mounted on the barrel arbor. When examining a typical eight-day longcase clock barrel, it will be noted that it has a spiral groove running across its surface giving usually between sixteen and eighteen grooves. The function of the groove is to feed the line across the barrel as it is wound. In practice, though, this groove is not essential. Many fine regulators do not have this groove, and in this instance, in the interest of the project being manageable in a typical workshop, the barrel will be smooth rather than grooved.
6. The winding ratchet wheel.
7. The front cap for the barrel. This has a hole drilled through in order to connect the line.
8. The click spring.
9. The maintaining power springs.
10. The click.

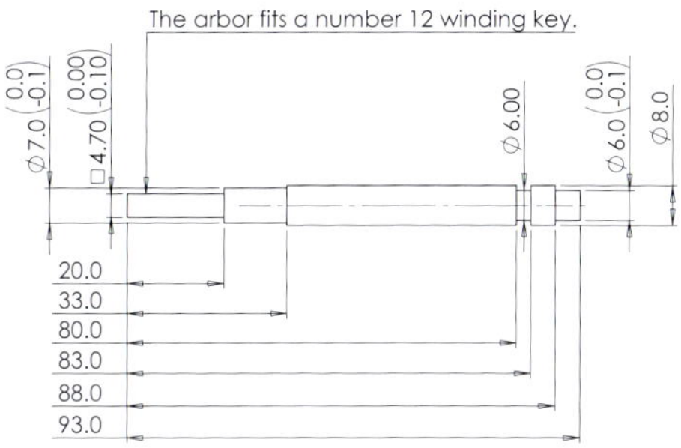

Key-wound barrel arbor.

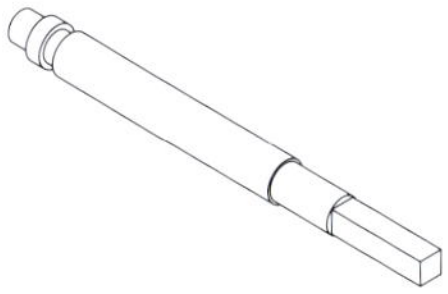

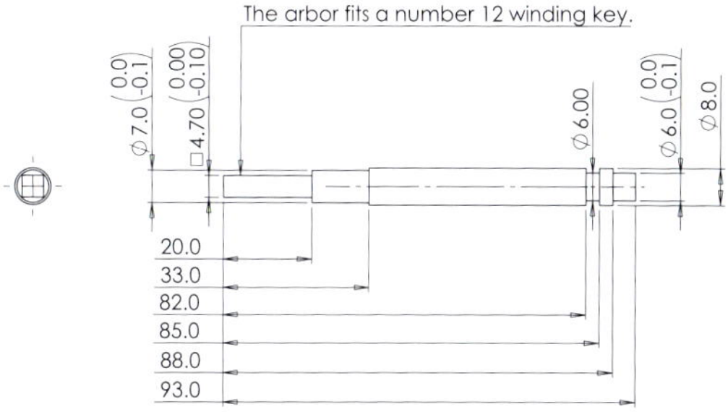

Key-wound barrel arbor using maintaining power.

Turn the arbor at quite a high speed and take small cuts.

Part the bar off at the total length of the barrel arbor using the parting tool.

The rear end of the arbor showing the rear pivot and circlip groove.

Return the bar to the lathe, turned end in, so the steel now ready to be turned is the end where the bar was parted. This end is now turned, forming the rear pivot, the rear end shake point, the circlip locating groove, the inner pivot for the great wheel. This allows the clock to be wound whilst the great wheel is engaged in the train.

The buff stick shown is being used to apply a basic finish to the arbor to make it more comfortable to handle. Although it cannot be seen in the photo the lathe bed is covered with disposable blue paper to ensure no debris from the buff stick falls onto the lathe bed.

Forming the winding square

Having turned the arbor, the next operation is to file the winding square. The question is whether to make the winding key and then the barrel arbor to fit it, or to form the winding square and make the key to fit? In this case the author's advice is to purchase a number 12 longcase clock crank key and make the barrel arbor square fit it. Filing a square into a blind hole is a tedious operation! Filing a square onto an arbor is a traditional apprentice-type exercise. If you are unsure of filing a square, proceed as described below but using an off-cut of steel bar as an exercise. It takes much less time to do a test piece than to make another complete arbor.

Take care not to overwork the file by using too much force, as it will clog and then score the surface. If the file continually clogs, try lubricating it with white spirit. Clear any file clogging as previously described. This is important, as it may not be possible to remove all the marks of a file-induced score and keep to a number 12 key.

Begin by turning a brass washer 8mm thick, with a hole 6mm diameter through the centre. An outside diameter of approximately 15mm is sufficient. This is the brass washer seen in the photos. Place the washer onto the winding square end of the arbor. The purpose of the washer is to protect what will be the end shake step for the front pivot from being accidentally caught by the file. Do take a minute to do this, as an arbor caught with a file is obviously a mistake and looks ugly.

Hold the arbor in the vice over the whole length being filed for the square. Taking the hand file with the safe edge running against the washer, file straight across the arbor. Make six file strokes across the arbor and examine it. There will be a line at the leading and trailing edge of the circle being filed, and these edges should be parallel. If not, wherever the lines appear to be further apart will indicate that you have filed with more pressure, causing the file to go more deeply in at this point. Take a couple more strokes to correct this. If it is not going well, place the arbor aside and do a practice piece. Patience pays here.

Assuming the described lines are parallel, turn the arbor 90 degrees in the vice, clamping the work on the filed side. In this way it can be seen the arbor is at 90 degrees to the starting point. File six file strokes and again check the lines. All being well they are parallel; turn the

The arbor clamped in the vice with the washer located. The first couple of file strokes have been taken, and the file has not been held flat, with more weight being put on the left-hand side of the file. This can be seen as the left-hand side flat section being wider than the right-hand side.

Too much pressure has been put on the right-hand side to correct the previous filing. Carrying out a practice exercise is worthwhile if you find you are proceeding in this fashion.

The aim is to have the lines parallel.

arbor so that it is now clamped on the newly filed surface. Again file six strokes, check and turn. In this way all four sides of the square will begin to be formed. Once the four sides of the square have been initially filed, remove the arbor from the vice and measure across the squares with the vernier gauge. They should be the same. Repeat the above process, beginning with where the vernier gauge reading was the greatest if there was a discrepancy. The vernier gauge will also indicate you are filing a square rather than an odd-shaped, four-sided shape.

Once the corners are starting to look like corners rather than bar stock, try the key and see how much further you have to file. It is a good idea to hold the square in the brass soft jaws from this point. It is a question of repeating this operation and checking often until the key is a tight fit on the square. (Do not get it stuck on the square!) This will allow enough steel to tidy the flats and the key to be a good fit. Finish the square initially by draw-filing with a fine file, and then use buff sticks, taking care not to round the flat surfaces as much as possible.

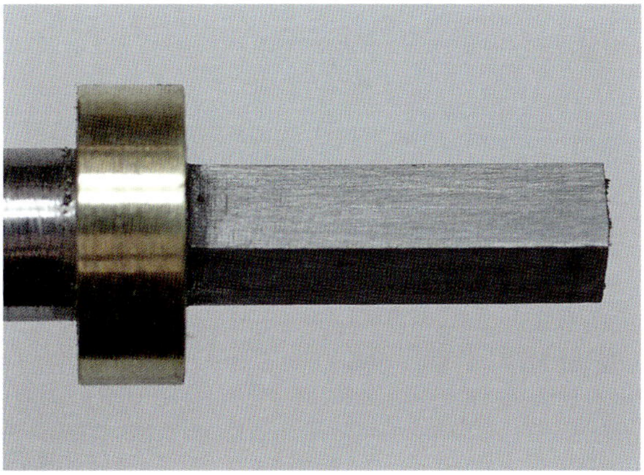

The square formed and draw-filed to finally fit before the final finish is applied to the square. If you have an old fine-cut file, a layer of the micro-finishing film can be stuck to it, and this used instead of a buff stick. This will help not to round off the flat areas.

FINISHING THE ARBOR

Having formed the square, the remainder of the arbor can now be finished. The pivots are polished in the lathe, using a very fine file to begin with. Do not use the pivot/burnisher file here, as it will not be wide enough and will very quickly clog. Once the pivots are smooth, work through the grades of buff stick to achieve a good polish. Once again, if you have an old smooth file, sticking finishing gels to this gives a very flat tool to finish the pivots. Finally, to finish the surface use a wooden buff stick with a cream polish.

Traditionally pivots are polished as the smaller-diameter pivots have been done with this project. Using a burnisher to polish the surface also hardens it, giving it a better running surface. Practically this is very difficult to do with the barrel arbor by hand. The use of a good-quality modern steel will enable the clock to run for many years without the arbor wearing unduly. This completes the barrel arbor.

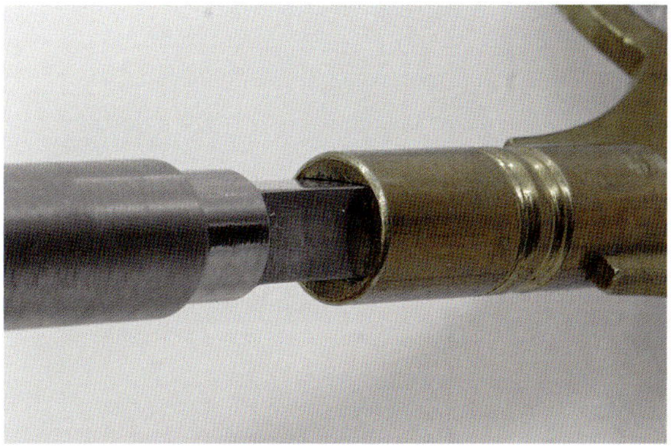

The square fitting the key.

The finished barrel arbor.

MAKING THE BRASS BARREL

For ease of supply and cost of brass tube the barrel was made with a length of brass tube of 50mm outside diameter, wall thickness 1.5mm, and 33mm (30mm when finished) long is required. This will need to be held in the chuck of the lathe to shape the ends. In order not to mark the barrel, either make up some soft jaw linings for the jaws of the chuck, as was done for the bench vice, or wrap the brass tube with a thin layer of brass or aluminium sheet.

Place the tube in the lathe and adjust it to run as centrally as possible. Carefully face the end of the tube. Often a burr is formed on the inside of the tube when facing the end; gently turn this away using the boring bar. Provided the tube is circular and of reasonable quality, it will not be necessary to turn the inside diameter of the tube. If the inside of the tube is uneven, use the boring bar and turn just enough away to even up the inner diameter of the tube to a depth of 4mm. Double-check the internal diameter of the tube, as you will need to turn the step on the end cap and ratchet wheel to be a firm fit into the barrel.

Traditionally, the barrel drum is pinned to the end cap and ratchet wheel with six steel pins on either end. The end caps are secured to the barrel arbor, and the assembly pinned together. However, it is not necessary to pin the barrel assembly together when using a brass cotton-reel design to mount the barrel to the arbor. The brass barrel shell will be held in position by the pressure applied to it when screwing the ratchet and end caps into place, provided the caps are a tight fit into the barrel shell.

As can be seen from the accompanying photographs, the author has turned the cap and barrel ends to be a secure push-fit into the barrel. In this instance the internal diameter of the brass barrel did not require turning. Drill a hole 1.9mm in diameter in the wall of the barrel as shown.

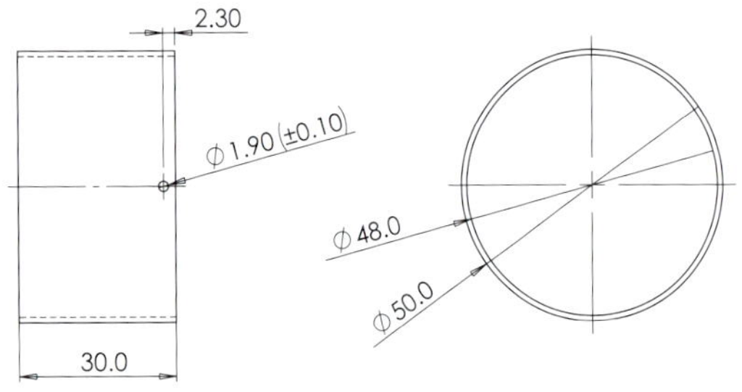

Barrel shell.

Fitting the ratchet wheel and end cap

Re-mount the ratchet wheel on the 8mm mandrel and place it in the lathe. The face of the ratchet wheel is now turned to form a step 4mm deep, which needs to fit snugly into the barrel section. Take care to turn the step into the correct side of the ratchet wheel, as it needs to wind in a specific direction. Note the drawing on the following page shows the side elevation and the front elevation looking at the raised area the great wheel will run against.

Next the blank end is dealt with.

The tube mounted in the three-jaw chuck with three brass soft jaws placed between the jaws of the chuck and the brass barrel shell. The end has been turned true.

The ratchet wheel held in the lathe and a step turned to fit snugly into the barrel shell.

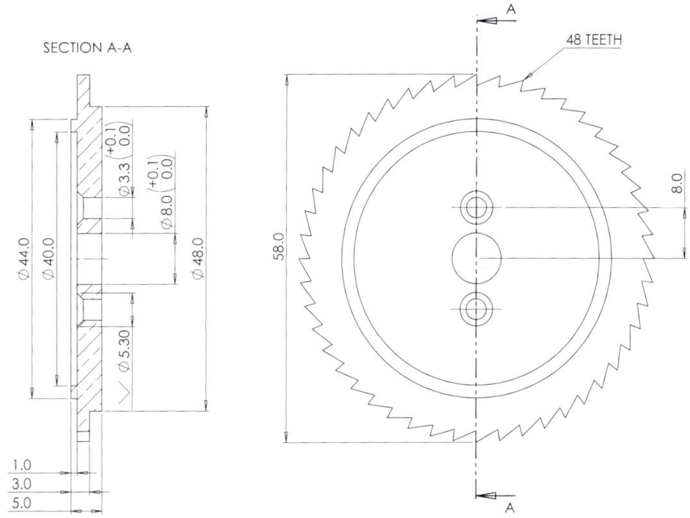

Winding ratchet wheel.

The end cap held in the lathe on the mandrel in order to turn the step to fit the cap onto the barrel.

Begin by effectively marking out another wheel blank of 54mm diameter. Drill the centre hole 8mm in diameter to fit the 8mm mandrel and drill a hole through at the place shown in the drawing to enable a bronze or gut line to be fitted in due course. Turn to the dimensions shown on the page overleaf.

Securing the barrel to the arbor

We now have a barrel, a ratchet wheel end cap and a blank end cap. Before assembling the barrel, a way of securing the barrel to the arbor is required. A straightforward way to achieve this is to turn a brass collar of 25mm diameter with a hole of 8mm to be a close fit onto the barrel arbor. It needs to be a snug fit. To locate the cotton reel on the arbor, measure in 18mm from the shoulder of the rear pivot. This is the locating point for the end of the cotton reel (*see* diagram). Secure this to the arbor with soft solder or Loctite 638.

The drawing overleaf shows the dimensions, and the photo shows the brass centre section – which the author has termed the 'cotton reel' – being turned in the lathe.

Now take the ratchet wheel cap and drill two holes as shown to enable tapping for two countersunk 4BA screws (of 3.0mm diameter). Place the drilled ratchet wheel onto the arbor against the brass cotton reel and,

The blank end of the barrel turned to fit snugly into the barrel shell. The mounting holes have been drilled.

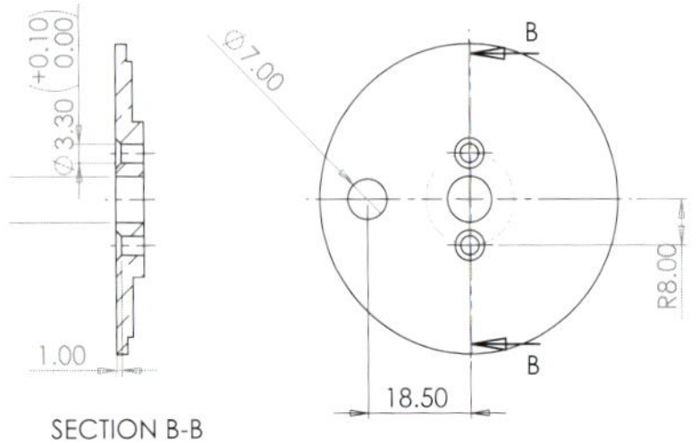

Front barrel cap.

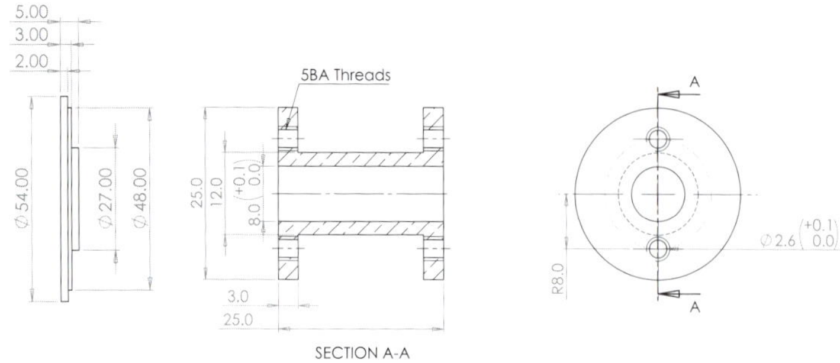

Centre of key-wound barrel.

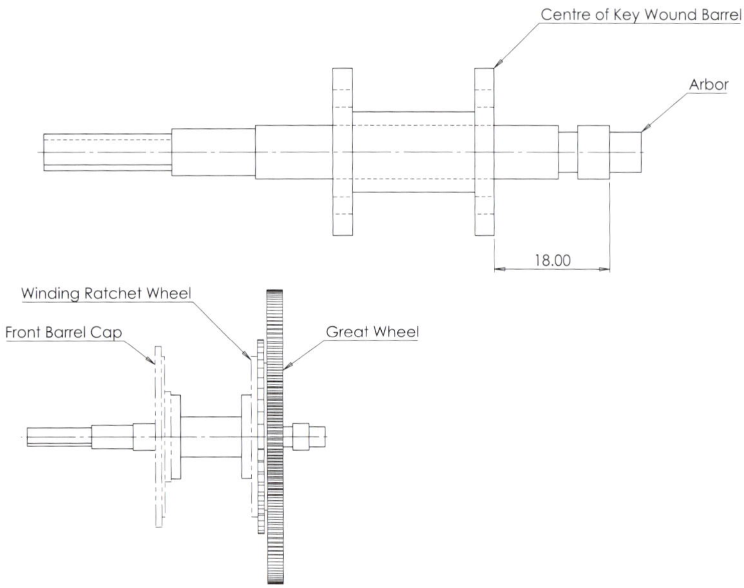

Centre of key-wound barrel position on arbor.

Beginning to turn the cotton reel.

The cotton reel attached to the barrel arbor, resting on the ratchet cap for the photo.

The finished cotton reel ready to be parted from the lathe.

Screw the ratchet wheel onto the brass collar with one screw.

using one of the drilled holes as a guide, drill through the cotton wheel. Now tap the hole in the cotton reel 4BA.

Open the diameter of the guide hole in the ratchet wheel to the diameter marked as 'clearance' hole in the Appendix – 3.7mm. Using this hole and thread, it is now possible to screw the ratchet wheel onto the brass collar with one screw. This locates the screw hole in exactly the place required, without the risk of mis-drilling and the hole not lining up.

With the ratchet wheel attached to the cotton reel, drill the remaining hole through using the 4BA tapping drill (diameter 3mm) and again open the hole in the cap, tap the thread into the cotton reel and you should find the second screw lines up correctly. Repeat the procedure with the end cap of the barrel.

The two ends of the barrel, with the centre tube between them, can now be screwed to the central cotton reel. This clamps the barrel into position without the need for further pinning or soldering.

Having assembled the barrel, the complete assembly is mounted in the lathe and the surface of the barrel carefully finished, using buff sticks and finally cream polish if you wish for a polished finish. Take care not to catch the tips of the ratchet wheel.

Maintaining power

Maintaining power is the term given to a device that keeps the clock running while being wound. If fitting the maintaining power option, the extra ratchet wheel needs to be fitted at this stage. If your clock does not include this device, move onto the section entitled 'The Winding Ratchet'.

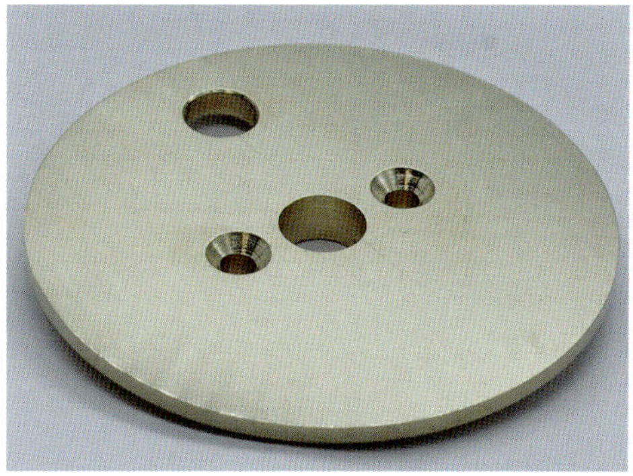

The blank end of the barrel with all four holes drilled ready to mount on the cotton reel.

The type of maintaining power shown here is referred to as 'Harrison' maintaining power, after John Harrison, the chronometer maker who designed it. A type known as 'planetary' is also found. Here we will make the Harrison type. The going train is not driven while being wound, since the driving force is removed whilst you wind the movement. The clock will lose time, and the train may even attempt to run in reverse during the winding process. For the recoil escapement this is not a concern, but for the dead-beat escapement it could damage the escape wheel teeth. When fitted, the winding click and spring are mounted on an extra ratchet wheel rather than the great wheel.

The following text and drawings describe the layout and dimensions for the ratchet wheel, tension spring, maintaining power arm (iron), its arbor, and click required.

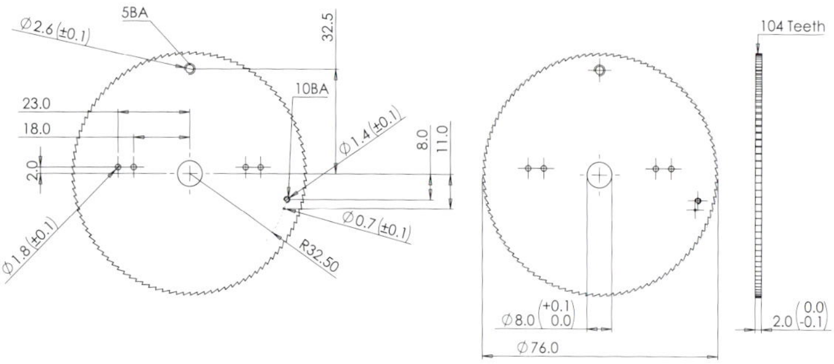

Maintaining power ratchet wheel.

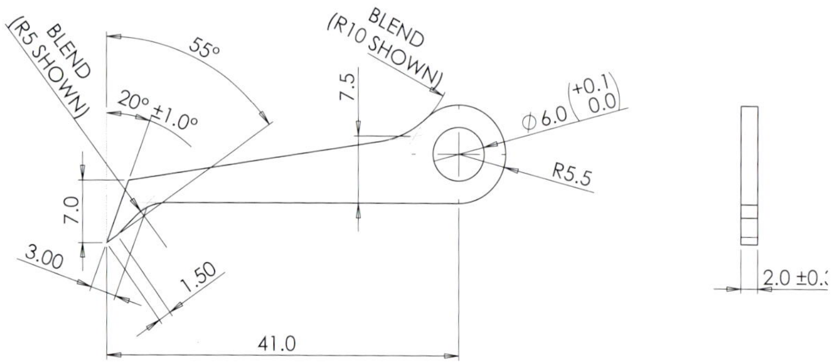

Maintaining power arm (iron).

The maintaining power ratchet wheel teeth being cut. Note the cutter is centred with the pointed edge lined up with the centre. The straight edge of the cutter is also therefore on the centre line.

The iron is marked-out on a piece of good-quality steel 2mm thick, using the scribe and rule as usual. The arbor can be turned from EN8 steel or pivot steel.

To construct these parts, the extra ratchet wheel is cut in exactly the same way as the winding ratchet: it is made from 2mm-thick brass, and does not require a rim to be turned into the edge, unlike the barrel ratchet wheel.

By now you will be familiar with the technique for marking and cutting the brass wheel blank, and turning the arbor with its collet for mounting the iron.

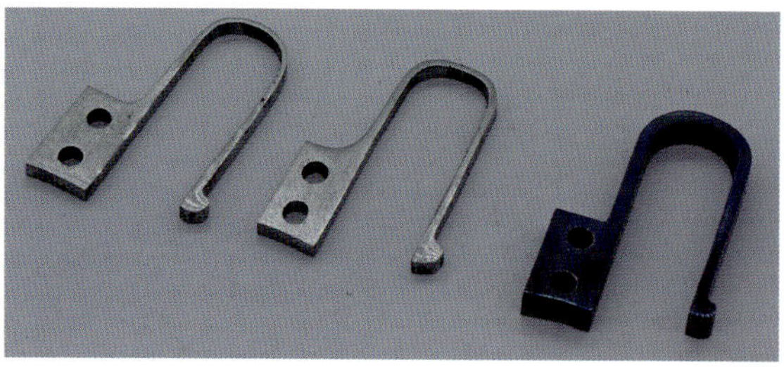

The shape of the maintaining power spring or springs. Two thinner springs give a more even force to the great wheel. The thinner springs require care when making.

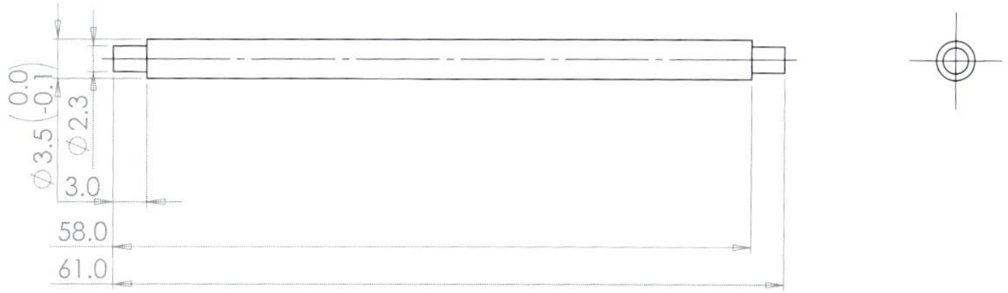

Maintaining power arm arbor 11-10

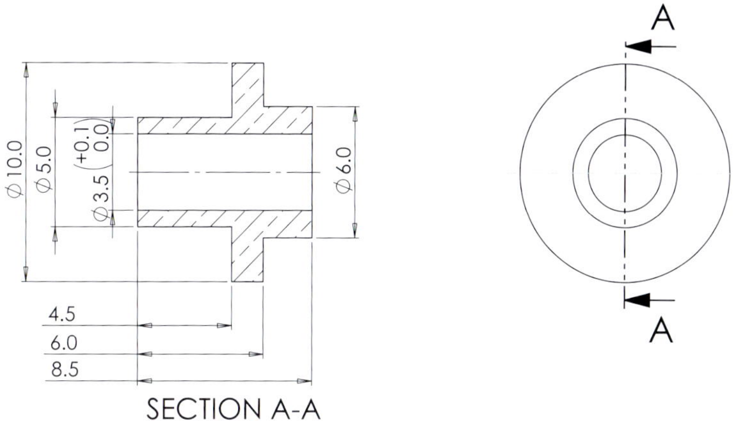

Maintaining power arm collet.

The two springs fitted to the barrel assembly. There is a photo in Chapter 3 of the single thicker maintaining power spring fitted.

All the dimensions are shown on the drawings. Take care to line up the iron on the arbor to locate positively in the maintaining power ratchet wheel. As with the train wheel, the iron-arbor requires a small amount of end shake.

The maintaining power spring or springs

The steel 'U'-shaped spring or springs, depending on your choice, provides the driving force to run the clock whilst being wound. Make this from a good-quality steel, as it will need to be hardened and tempered correctly to give an adequate force without being too brittle – which would cause it to snap. The author has used precision-ground flat stock steel. Using the piercing saw, cut the shape out, drill the mounting holes, and finish the spring *before* hardening and tempering it. The finishing is applied at this point even though it will be finally finished again. This is because the steel is much easier to finish before it is hardened and tempered. Either make one spring that is a little thicker or two thinner springs. The thinner ones will give a gentler force, tend to work better, and spread the force over both sides of the great wheel.

Hardening and tempering the spring

To harden the spring, it will need to be heated to red hot. There are many shades of red hot. These range from a dull red glow to a bright cherry red, before the steel begins to fizz and spark, at which point it is burning and melting. The steel only needs to be hot enough, and not overheated. For this process, the steel used by the author to make the spring only needed to be heated to a dull red glow. Heat the spring evenly

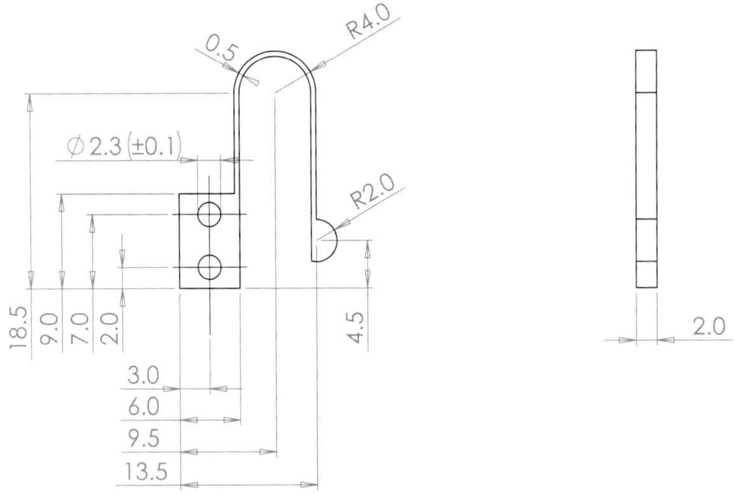

Maintaining power spring.

The roughly cut spring prior to drilling the screw holes and filing to shape.

so that the mounting section of the spring turns red as the thinner section turns red.

As you heat the spring the steel will change colour. Pay attention to this as it becomes important in due course. The first change is that it looks more polished. Next, it turns a light straw colour; this darkens until it turns purple and then quickly blue, before going a dull grey colour, followed finally by red hot. The red becomes brighter as more heat is applied.

Once the spring is glowing with a dull red colour, plunge it into a pot of cold water. Oil is sometimes used instead of water; use water here. The spring will now be hard and very brittle. If you were to try to flex it, it would snap. It will look dull and have a flaking layer known as fire skin on the surface.

To temper the spring, carefully clean the fire skin away from the top of the spring and place it either on a brass plate or a tray of brass filings. Heat the brass plate or the underside of the tray of filings in order to heat the spring until it begins to turn a straw colour. Watch carefully: when it turns a slightly darker straw colour, it will be close to turning blue.

For this project the author found the steel used in the photographs needed to be tempered to blue as it snapped when left straw-coloured. You may find the steel you use becoming too soft when tempered to blue. If this is the case, reheat it to red, quench, tidy and then heat to dark straw colour.

An alternative to making this spring is to purchase two ready-made springs sold as click springs from the horological supplier. These work well; two will be required.

THE WINDING RATCHET

The click is the name given to the steel ratchet hook that locates in the ratchet wheel teeth. It is operated when the clock is wound, with a spring keeping it located in the teeth of the ratchet wheel to stop the clock from immediately unwinding when you release the key.

Begin by crossing out the great wheel to one of the drawings shown in Chapter 7, depending on whether you are crossing the wheel with the traditional pattern, or with six crossings, with or without fitting the maintaining power device, to the pattern shown.

The section left over in two of the corners is for locating the click. Note that these are not required if fitting

An assembly made by the author for a version of the deadbeat escapement movement using a smaller diameter barrel and two supplied click springs acting as the maintaining power drive springs.

To complete the barrel assembly as a whole, we have a click.

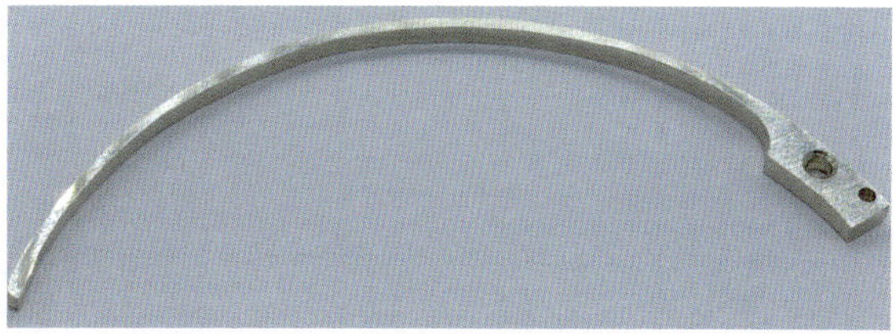

Click spring.

the maintaining power assembly. The rim of this wheel is left thicker than the wheels further up the train for strength purposes, and also to give sufficient thickness to mount a click spring. (If fitting maintaining power, the click and spring are mounted on the maintaining ratchet wheel and not the great wheel.)

For this movement the click is a flat piece of steel, 2mm thick, shaped to accommodate a countersunk shouldered screw to secure it in place. A countersunk screw is used so the screw head does not interfere with the line as the clock is wound. Mark out the

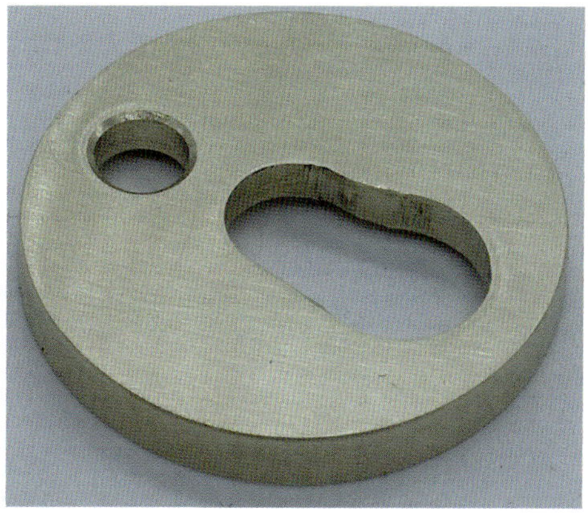

Securing clip – a circlip.

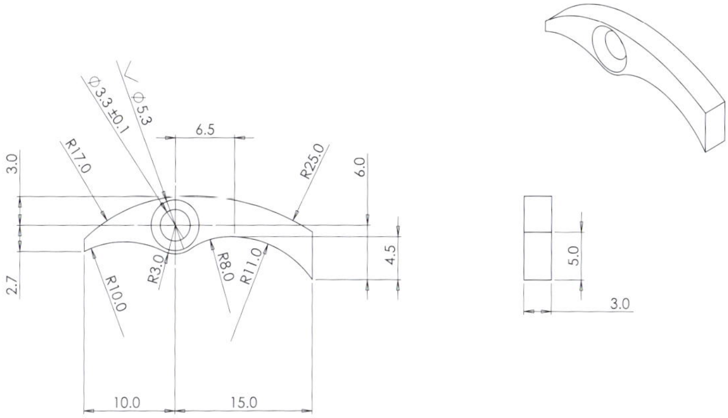

Key-wound click.

click as shown in the drawing, drill a hole of 3.4mm diameter, and countersink this hole with a commercial countersunk bit. Using the piercing saw, cut the click out, and file the edges to tidy the shape. Finish off by draw-filing the edges, and polishing with wet and dry paper from 240 grade to 2,000 grade if you would like the edge polished. For operation the edges need to be smooth, but not necessarily polished.

Making the screw is a straightforward turning exercise, follow the dimensions shown.

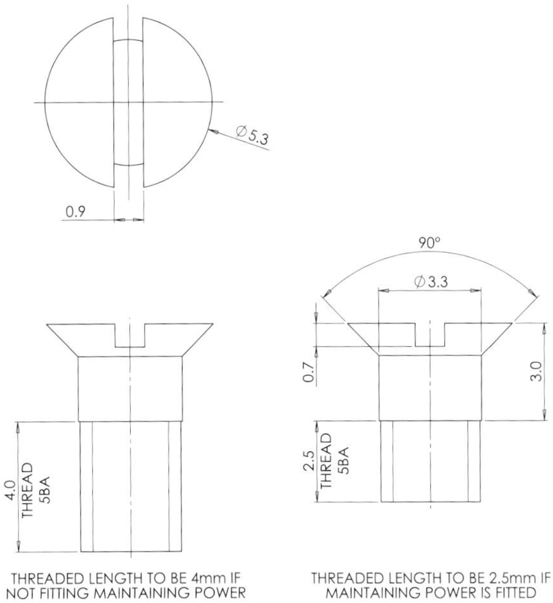

Click screw.

THE CLICK SPRING

To make the click spring we firstly need to make the brass suitably 'springy', if you will. Brass cannot be hardened by heating and quenching. The only way to harden brass is to push the molecules together either by using a roller or, as in this instance, by beating the surface with a hammer.

Take a piece of approximately 3mm-thick brass sheet (more than the 2mm we want to end up with and larger than the spring) and place it on the anvil. There will also be the circlip to make, which can either be made from hammered brass, or turned from solid brass bar. The procedure for making the circlip follows later in the text.

There is a handy traditional tool called a hand vice that is useful here. The brass now needs to be held on the surface of the anvil and beaten with the hammer. The hand vice will keep your fingers further from the hammer; if you do not have a hand vice, use a slightly larger piece of brass to keep your hand clear of the hammer. Taking care, use the smooth end of the hammer to beat the brass. Try not to hit it very hard with the corner of the hammer head as this will put deep dents into the surface – these dents will need to be removed from the finished spring. This does take a little time; hammer both sides of the brass to keep it flat. Next, mark onto the surface of the brass the shape of the spring to be made. Centre-punch the holes shown in the drawing; these will be the holes to accommodate the screws for securing the spring to the great wheel.

As with the click, having cut the spring out, file the spring finally to shape and draw-file the cut edges to tidy them. Take care not to kink the spring, as it will be likely to crack when you try to straighten it again. (Even if it does not crack, it will have a weak point, which will

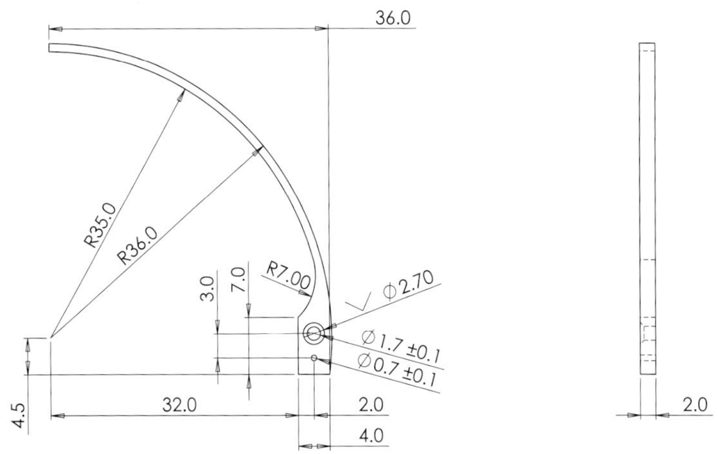

Click spring.

Using a hammer and anvil to work-harden the brass.

fail in use.) Rub the spring onto the surface of the finishing films, and using water to lubricate, tidy up the facing surface. The underside need not be finished to any great extent – just enough to make sure there are no high spots that could catch the surface of the great wheel and cause the spring to operate incorrectly.

Fitting the click and spring to the great wheel

If your choice is not to fit maintaining power, having made the click and spring, the next task is to mount them onto the rim of the great wheel. Mount the click first, by drilling a hole of 2.6mm diameter through the crossing corner – where extra brass was left in place to accommodate this – to allow a thread of 5BA to be tapped through, matching the shouldered screw. Screw the click into place: it should move freely, but not be loose. If it is a bit tight, free it by gently deepening the countersink in the screw hole a very small amount and trying it again. If the fit is too loose, turn the height of the shoulder of the screw down a small amount and retry. If the screw cannot be satisfactorily held in the lathe to reduce the shoulder, it is best to make a new screw. The location of the click spring can now be decided.

Place the spring onto the surface so that the end acts on the click as shown overleaf. Using a scribe, mark through one of the screw holes. Remove the spring, centre-punch the mark just made, put the spring back on the great wheel rim and check the centre mark is where you intended it to be before going ahead and drilling a hole of 1.4mm diameter to allow a thread of 10BA to be

The brass once hardened.

tapped through. Fit the spring to the great wheel with one screw; this will enable you to mark the correct location for the second hole to accommodate either a second 10BA screw, or a steel steady pin. If you use a steady pin, drill this hole 1mm in diameter and fit a steel pin through the hole in the spring passing though the great wheel.

You can try a trial turn and see that the ratchet operates correctly. The sound made is the sound you will hear every time you wind the clock. Progress is being made!

Fitting the click and spring to the maintaining power wheel

If your clock has the maintaining power option, the procedure differs. The click and spring are fitted to the maintaining power ratchet wheel as shown in the diagram on page 104. The only difference in constructing the click, spring and screw is the thread cut onto the click shouldered screw is 2mm rather than 4mm in length. *See* page 104 for the dimensions for locating the click, spring, and maintaining power springs.

Place the maintaining power ratchet wheel against the ratchet end of the barrel and place the click in position so it acts into the teeth of the ratchet wheel. Mark the centre of the hole by using the drill used to drill through the click mounted in a pin vice, so it can be gently rotated to mark the surface of the brass. Centre-punch and drill this hole 2.6mm diameter to enable tapping to 5BA. Fit the click with its screw and check it will rotate freely without being loose. If not, follow the procedure stated above for adjusting and fitting the click. Provided all is well, the spring can now be located and the centres for the holes marked in the same way as noted for fitting to the great wheel. Drill and tap these for 10BA and screw the spring into place.

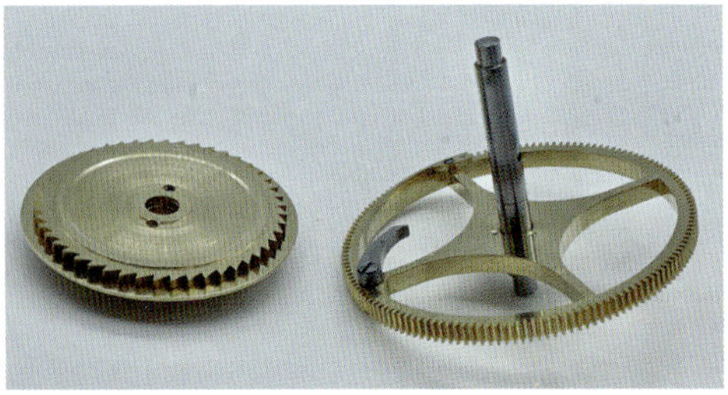

The click positioned and the location of the click spring acting on the top of the click. (Shown is the chain-wound barrel.)

Place the great wheel against the ratchet wheel with the spring and click attached, locating the click into the teeth of the ratchet.

The bar turned to the correct diameter. Whilst in the lathe, the central hole of the diameter shown in the drawing can be drilled. This is the hole the clip uses to rotate around the arbor, so it needs a smooth surface. A marking-out line to help locate the clearance hole for the barrel arbor to pass through can be drawn using the lathe tool before parting the disc from the bar.

The centre point for this hole has been centre-punched ready to either drill or mark out an 8mm-diameter hole.

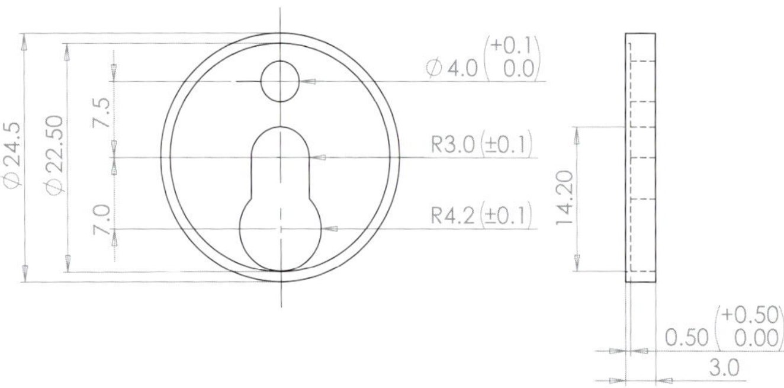

Circlip.

THE CIRCLIP

The last part of the operation is to make the securing clip holding the whole barrel assembly together.

If making this from the beaten brass sheet, make this clip a little domed to give a small amount of tension to the finished assembly. A trick here is to use the hammer and beat the centre of the brass; it will form a dome as the brass is stretched with the hammering. It is not so easy to mark-out or saw the brass once domed, but it can be done with a little patience. To help with the marking-out you could draw out the shape required and stick to the surface of the brass with double-sided tape, and then use this as a pattern. Should the result not prove satisfactory, you can cut the clip from flat brass and bend it to produce a small amount of tension.

Alternatively, a flat circlip of the correct thickness can be produced. Either use 2mm-thick brass sheet or turn a section of brass bar to the correct diameter in the lathe, as shown.

Whichever method of construction you use, the edges of the inside holes and the outer edge need to be finished in the usual way. The facing surface of the clip can be finished using buff sticks.

To finally fit the clip, firstly try it on the barrel arbor without the great wheel in place to see if it fits into the barrel arbor slot previously machined. If the slot is a bit tight, ease it with a needle file – but remember, it has to be a close fit.

The whole assembly can now be put together for a trial fitting. Once the clip is in place the great wheel should turn smoothly – but not loosely – round, whilst being held against the ratchet wheel. If it is very tight and you made a domed clip, remove the clip and very gently reduce its height by placing it in the vice and gently applying pressure to reduce the height. It is important not to over-do this as you may crack the brass, and also it is not easy to increase the height again. If you made a flat clip it is a question of gently thinning the clip by using the abrasive gel sheets until it fits.

MAKING THE CHAIN-WOUND BARREL

Whilst not actually a barrel, the term is used here for ease. The chain-wound barrel is more straightforward to construct, having fewer components than the key-wound variant. Bear in mind you will need to place the finished movement higher up as the weight falls more quickly. (This is why older longcase clocks that are chain- or rope-wound tend to only operate for 30 hours as a striking clock.)

In order to ensure a ready supply of suitable chain for this project, one has been selected with the same dimensions as those used for modern German weight-driven clocks manufactured by the Hermle company.

MAKING THE ARBOR

Mount a piece of EN8 or similar steel bar of 14mm diameter in the lathe and turn the barrel arbor to the dimensions given. This arbor is more straightforward to turn than the key-wound barrel arbor. The pivots are polished in the lathe, using a very fine file to begin with. Do not use the pivot/burnisher file here, as it will not be wide enough and will very quickly clog. Once the pivots are smooth, if you have an old smooth file, sticking finishing film to it will make a very good buff stick that will polish the pivots. Work through the gels to the very fine grade. The file covered in the gel sheets gives a very flat tool to finish the pivots. To finally finish the surface, use the green gel sheet and then a leather-covered buff stick with a cream polish. Traditionally, pivots are polished as the smaller diameter pivots have been done with this project. Using a burnisher to polish the surface also hardens it, giving it a better running surface. Practically this is very difficult to do with the barrel arbor by hand. The use of a good-quality modern steel will enable the clock to run for many years without the arbor wearing unduly. Note the four holes to be drilled into the arbor. The hole of 2.8mm diameter located 21.5mm from the arbor end is to enable a 6BA screw to pass through the arbor and screw into a brass collet. Do not drill this hole until it is noted on page 117.

The remaining three holes are in the 14mm diameter step and drilled on a diameter of 12.5mm, these enable the great wheel to be riveted in place using three steel pins shown on page 113.

The great wheel could also be soft-soldered into place, or a modern adhesive such as Loctite 638 could be employed. The great wheel needs to be a snug fit to this step in order to run true and for the riveting pins to hold securely.

As can be seen, the great wheel is fixed to the arbor for this type of winding work, with the click and click spring mounted in the same way as the key-wound barrel. The ratchet wheel and winding sprocket are then set to operate against the great wheel, being held against it by a

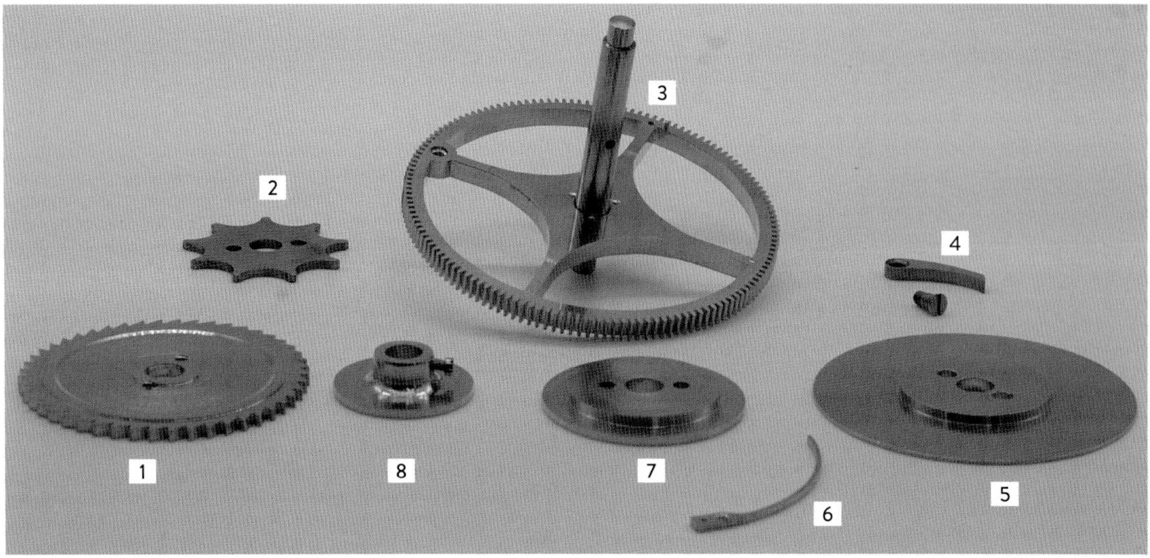

The components for the chain-wound barrel: 1: winding ratchet; 2: chain sprocket; 3: great wheel mounted on the arbor; 4: click and screw; 5: chain rear guide; 6: click spring; 7: front chain guide; 8: securing collar.

brass disc in this instance. The traditional 30-hour English longcase clocks have a slightly different arrangement to the one described here. However, using this method gives much smoother winding and avoids the crossings of the great wheel becoming worn over time. This also allows the great wheel to be crossed with six crossings if desired.

The great wheel is riveted to the long end of the arbor using three steel pins for rivets. Place the great wheel on the arbor (use soft brass jaws in the vice) and then wheel-down in the vice. The holes in the steel step can be used as a drilling guide; drill right through the great wheel. Remove from the vice and insert three tapered pins; rivet these over, and try to make the rivets as tidy as possible. Using a flat-ended punch here helps to avoid catching the steel collar with the hammer. If necessary, return the assembly to the lathe and carefully turn the top of the rivet so it is smooth and almost flush with the wheel.

The remaining parts to be made are the ratchet/sprocket assembly, and the securing collar with its locating screw. The ratchet wheel pulley assembly rotates on the arbor against the great wheel. The ratchet wheel is the base to which the components of the pulley are secured. It has two threaded holes, used to secure the sides of the sprocket and the steel sprocket together. Turn the ratchet wheel to the dimensions shown, taking care to turn the correct side. The dimensions for the ratchet wheel and click spring are the same as for the key-wound barrel. Note the click itself is a different shape to that used for the key-wound assembly.

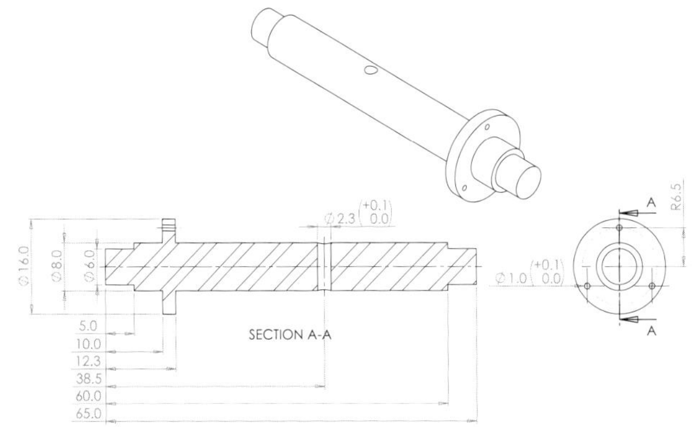

Chain-wound barrel arbor.

The sprocket is made from steel, as this will wear much better than brass. The spacing of the teeth has to match the chain. You might find it best to use the drawing as a pattern. Copy the pattern – double-check the dimensions of your copy – stick this to the steel, centre-punch all the centre holes, and cut the sprocket out using the piercing saw.

To save some sawing and finishing, the drawing shows the centre of each of the nine semi-circles used to form the sprocket (*see* page 115). Using a centre punch, mark each point and drill each hole to a diameter of 12mm. Begin by using a centre drill to ensure the actual centre is drilled. The author drilled the centre hole of 8mm diameter into the uncut sprocket, enabling the sprocket to be mounted on the 8mm mandrel. A rotary table was secured to the pillar drill table in order to rotate the sprocket. This allowed the author to drill the holes in turn working through a series of drills to the diameter of 12mm. (A rotary table is not essential to make the sprocket.) Use oil to lubricate the drills. All that is then left to do to acquire the basic shape is saw the tips of the teeth with the piercing saw. If you have not done so, drill the centre hole to 8mm diameter to fit the barrel arbor, and the two mounting holes at 3.7mm diameter to allow the mounting screws to pass through in due course.

The photos show using a rotary table – in this instance a CNC table – to accurately drill the nine holes required to form the chain sprocket.

Once the sprocket is cut, mount it in the lathe on the 8mm mandrel. Carefully turn the diameter of the sprocket until all the teeth are the same height. This takes care of any discrepancy in marking and sawing the teeth of the sprocket.

The rear end of the barrel arbor turned.

The arbor marked and centre-punched ready to drill the holes to rivet the great wheel to the arbor.

The holes drilled in the barrel arbor step, ready to drill through into the great wheel.

Next, adjust the angle of the lathe tool to allow a chamfer to be turned on the tips of the sprocket teeth.

With the **lathe turned off**, try the chain on the teeth to ensure the links allow it to slide down the sprocket teeth. DO NOT under any circumstances turn the lathe on! Finish with buff sticks to form a smooth surface so that the brass chain can smoothly engage with the sprocket. Make sure you check all the teeth – they must engage *and* disengage cleanly. If there is a tendency for the chain to hang on a tooth, carefully file the tooth concerned to smooth the action. Use a fine file to make any adjustment required. (*See* page 116.)

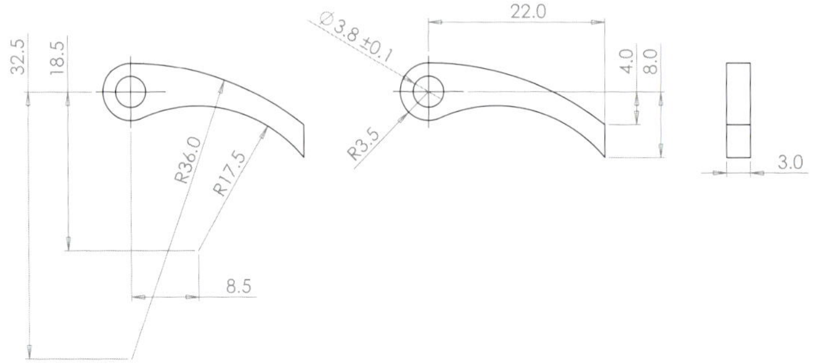

Key-wound click.

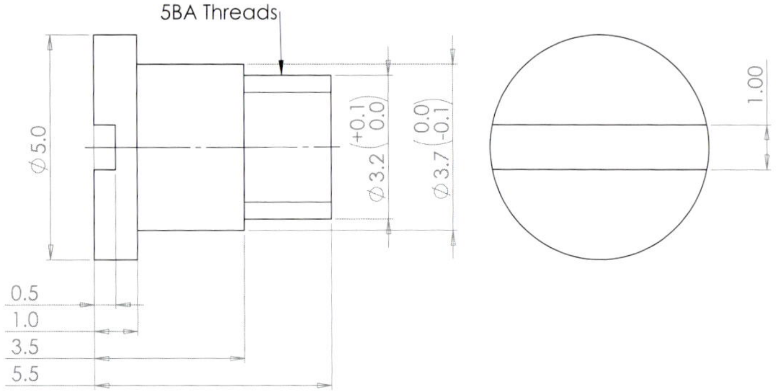

Shouldered screw.

A clock movement can be an unforgiving mechanism. Any odd things left in construction – such as a tight pivot or a chain sprocket tooth that catches the chain – will always stop the clock. One never 'gets away with it', so to speak.

The steel sprocket teeth engage with the chain to stop it slipping. The two brass plates either side of the steel sprocket help to turn the chain as it comes into the sprocket assembly to ensure the holes in the chain links engage with the sprocket. The larger-diameter disc shields the winding ratchet and click from the chain.

To construct the side plates begin by making the blank discs as you would make a wheel blank. Drill the centre hole of the two side plates to a diameter of 8mm and mount them one at a time on the mandrel in the lathe. Turn down each to the dimensions shown on the drawing. While each side plate is mounted in the lathe, it is a good opportunity to mark-out for the two screws used to assemble the complete winding sprocket. Take care here, as the holes will need to line up correctly.

Mark a circle of 32mm diameter using the lathe tool, and on this line mark two divisions 180 degrees apart. These are marked by passing the lathe tool across the surface. (Lathe off!) Drill through the holes in the rear larger side plate using a 3mm-diameter drill. Place the winding ratchet and the larger side plate together on the 8mm mandrel and, using the holes as a guide, drill through the ratchet wheel. This gives an accurate location for the screw holes. Tap these holes with the 4BA tap. Next, drill through the holes in both the side plates, opening up the holes in the larger one, to 3.7mm. Use countersunk screws so that the head of

The click marked-out.

The click cut out and ready to fit.

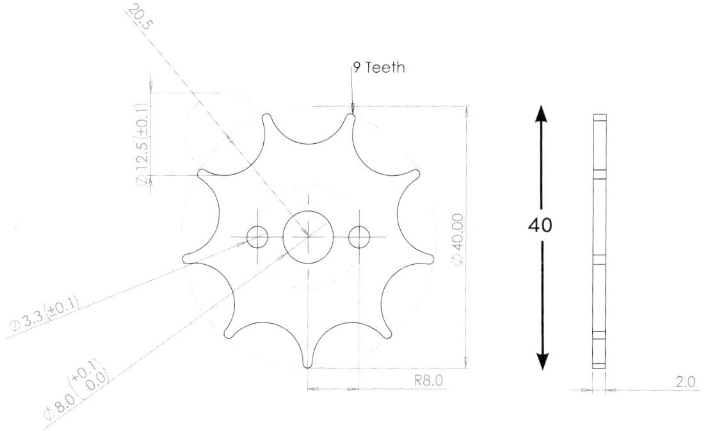

Sprocket for chain-wound barrel.

Use a centre drill to make the first hole ready to guide the next size drill.

The holes are drilled to 12mm diameter and the tips cut through.

the screw sits flush with the surface of the side plate, and does not interfere with the operation of winding the clock.

It will now be possible to secure the four parts of the sprocket together using two 4BA steel countersunk screws. The screw threads may need to be shortened so they do not stick out beyond the outside of the ratchet wheel.

Screwing the assembly together rather than riveting it allows it to be dismantled for cleaning. Provided the correct chain is used, it will pass through the sprocket without snagging or slipping.

SECURING COLLAR

You should now have a complete winding assembly to test; all that remains is to secure it to the great wheel. Traditionally a spring and clip arrangement is used to keep the sprocket pushed hard against the crossings of the great wheel. In effect, the crossings were used

The first set of holes drilled to diameter 6mm.

Once the sprocket is cut, mount it in the lathe on the 8mm mandrel. Carefully turn the diameter of the sprocket until all the teeth are the same height. This takes care of any discrepancy in marking and sawing the teeth of the sprocket.

Adjust the angle of the lathe tool to allow a chamfer to be turned on the tips of the sprocket teeth.

Use a fine file to make any adjustment required.

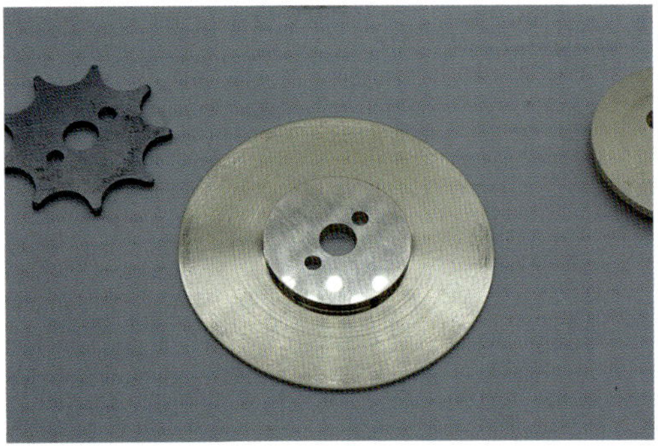

This large diameter disc shields the winding ratchet and click from the chain.

This again is a turning exercise. The author has used a round-tipped tool of diameter 5mm to turn the smaller diameter of the collar in order to give a neat curved corner at the point the diameter increases.

as the ratchet wheel teeth. Instead of a smooth constant clicking heard when winding, as we have here, a rather lumpy clunking is heard. Over time this wears the crossing, so here we are using a ratchet wheel and collar.

Turn the collar to the dimensions shown. The central hole diameter is drilled to match the arbor. It is best to measure the arbor you have turned, and drill the collar to match this closely (8mm in the one photographed). It needs to slide along to meet the sprocket. Drill a hole of 1.8mm diameter into the collar 3mm in from the end. This will give the correct size hole to use as a guide for an 8BA screw. Fit the collar hard against the sprocket and use the same drill to mark the position of this hole on the arbor. Remove the collar, centre-punch and drill the hole 1.8mm diameter and tap with an 8BA tap. Screw the collar in place and try turning the sprocket. It will probably be a bit tight: if too tight, turn a very small amount from the larger-diameter end of the collar and re-fit. When the clock is finally assembled, there will be a small amount of grease added to the underside of the collar, which will help with the smoothness of operation.

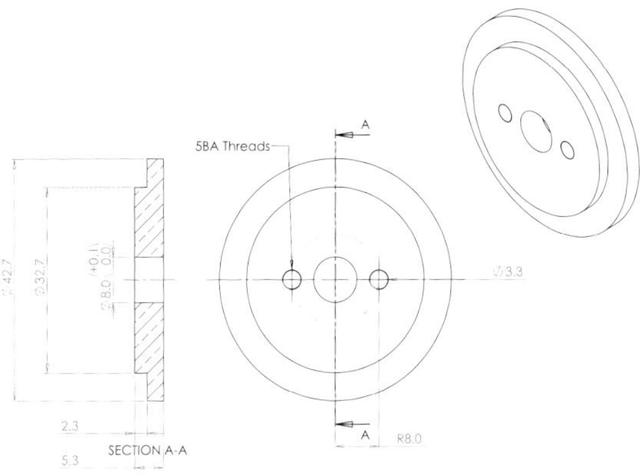

Chain-drive-sprocket front guide.

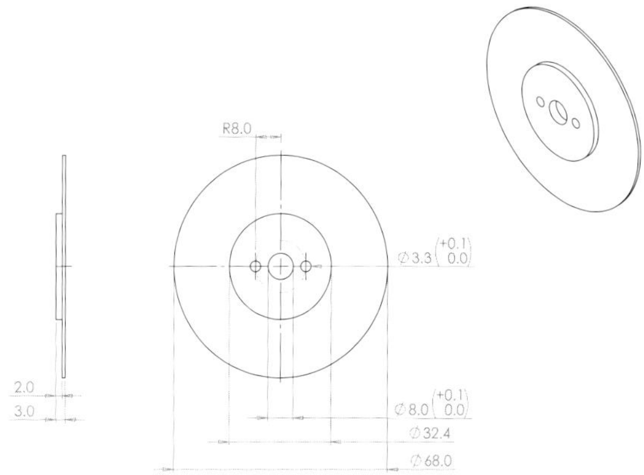

Chain-drive guide.

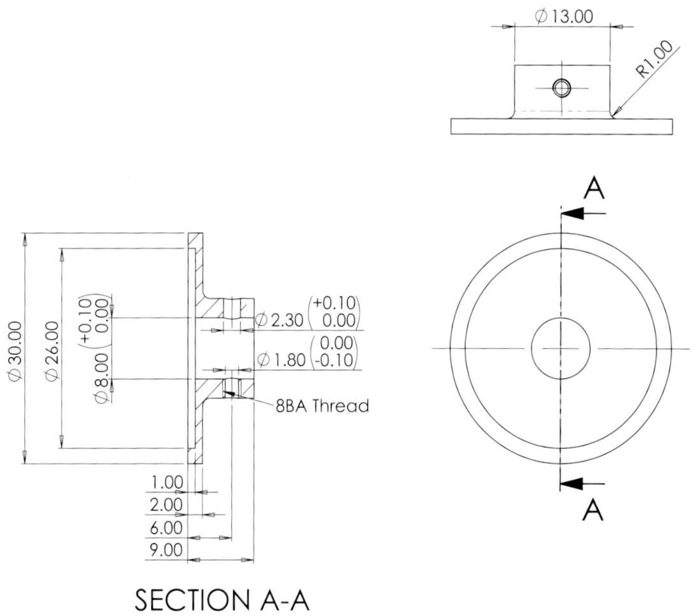

Securing collar for chain-wound barrel.

Setting the Train into the Plates

10

THIS IS THE POINT at which the clock movement begins to really take shape and the previous work all now seems worth the effort! Note that the same set of plates was used to mount both clock movement variants. You may find there is an empty hole for the locations of the maintaining power arm and both winding variants in the accompanying photos.

Having made and mounted the wheels and pinions together, along with your chosen barrel assembly, the next operation is to set them in the correct location within the plates.

The plates joined using the brass threaded plugs.

The recoil escape wheel train and the chain-wound barrel mounted between the plates. (Note that the same set of plates was used to mount both clock movement variants, so there is an empty hole for both the location of the maintaining power arm and the key-wound barrel.)

A clock movement is an unusual piece of machinery in that whilst it is quite precise, it is not like an internal combustion engine or similar object. There have to be some built-in tolerances to allow for the movement to be free running. The two tolerances of concern to us as we come to fit the clock wheels and pinions into the plates are the tightness of the pivot holes around the pivot, and the ability for the arbors to 'float' between the plates. This float is known as 'end shake', and is the term we will use from now on. The subject of end shake and how to adjust it is dealt with towards the end of this chapter.

To begin, it is important to be able to drill at least a pilot hole marking the centre for each pivot through *both* plates at the same time to make sure the holes really line up well. Remove the pillars from the plates and refit the 4BA tapped brass plugs made for shaping the plates, joining the plates together as they were when the edges were finished.

Before dealing with where the holes will actually be located, the following notes describe how the pivot holes are drilled and the pivots finally fitted.

DRILLING THE INITIAL PIVOT HOLES

The pivot holes are not initially drilled through to their final diameter. The final diameter is cut using a five-sided clockmaker's broach. This allows for any discrepancy you may have in your work and to achieve the best final fitting, with a smooth-sided, slightly tapered pivot hole.

Generally, High Speed Steel (HSS) twist drills are available in steps of 0.1mm. Remember we are drilling through both plates at the same time to ensure the pivot holes line up correctly. Make sure you measure the diameter of each of your finished pivots for each arbor, and work with the *smallest* diameter on each arbor for the next operation.

Drill your guide hole 0.1mm smaller than the smallest pivot diameter. For example, if the rear centre wheel pivot measures 1.6mm in diameter, drill the hole 0.1mm smaller at 1.5mm. The diameter of the front centre arbor is larger; this can be drilled to within 0.1mm *once the plates have been separated.* All the pivot holes are finally fitted using a five-sided clockmaker's broach other than the barrel arbor holes. These will be drilled to the actual size, as this is more practical with the larger-diameter pivots.

USING A FIVE-SIDED BROACH

The procedure for using a five-sided broach is to begin by selecting a broach that passes through the hole approximately a quarter of its length.

Shown is the five-sided broach passing through the pivot hole in the plate, from top to bottom in the photo, from the inside of the plate to the outside.

Put the broach into the hole from the inside of the plate, pointing towards the outside. Next, taking care to hold the broach and making sure it is straight in both directions, gently rotate the broach and it will cut away the inside of the hole. Cut just a small amount and try the pivot – take care not to get the pivot stuck in the hole! Repeat these two steps until the pivot will pass through the hole.

The question is when is the pivot the correct fit and when is it too loose? This is one of those things experience at the workbench teaches you, but as a guide here, when the arbor will move from the vertical just a degree or so, it will be a good fit.

Do not force the broach into the hole as it will get stuck, but rather once the broach comes to a stop as you pass it into the hole begin to turn it. Five-sided

Here the rear centre arbor pivot is fitted into the pivot hole and a set square placed against it to give an indication as to how well the pivot would ideally fit the pivot hole.

broaches cut by turning in either direction. How much to cut is not an exact measurement but a 'feel' that is learnt over time. Provided the pivots are finished well, and the wheels made and depthed correctly, very little 'play' will be required in the pivot holes for the clock to run freely.

This is the way pivot holes were traditionally fitted, and we still use the method today as it works well.

Having used the five-sided broach to open the pivot hole to the correct diameter, there will be a small brass burr around the edge of the hole. This is removed by using a flat countersink.

Note: The front centre and both barrel arbor pivot holes are best drilled through to the correct diameter with a drill. Provided the diameter of your pivot is accurately turned it would be best to finish the hole using an engineer's reamer of correct size – as was used for the pillar plate holes. Larger five-sided broaches are available. Care is required when using them, especially as applying too much pressure when opening out the diameter of the hole can result in the hole having five 'sides' and five 'points' which looks odd, and will not give a satisfactory bearing surface.

MOUNTING THE CENTRE AND THIRD WHEELS

Whether constructing the deadbeat or recoil escapement version of the movement, the first two arbors to be mounted in the plates will be the centre and escape wheel assemblies. Their location can be taken from the accompanying drawing.

Note that the location of the escape wheel differs depending on which escapement you have chosen to make.

These are located first because the remaining train wheel, the third wheel, is depthed to run correctly with both the escape wheel pinion and centre wheel using the depthing tool.

The centre arbor carrying the hands is generally set to pass through the middle of the dial. We also know the location for the escape wheel, which, if you are planning to fit a seconds hand to your movement – as the author chose to do – will show in the correct location for the seconds scale to be mounted on the dial. The escape wheel must also be located to allow for the pallet bridge to be mounted in the correct place for the pallets to function, and the pendulum to be supported.

The location of the centre and escape wheel arbors is found by measuring up the centre line from the bottom of the plate to the dimensions shown on the drawing, marking both these points with the centre punch. Drill through both plates using a drill 0.1mm smaller in diameter than the *smaller* pivot diameters.

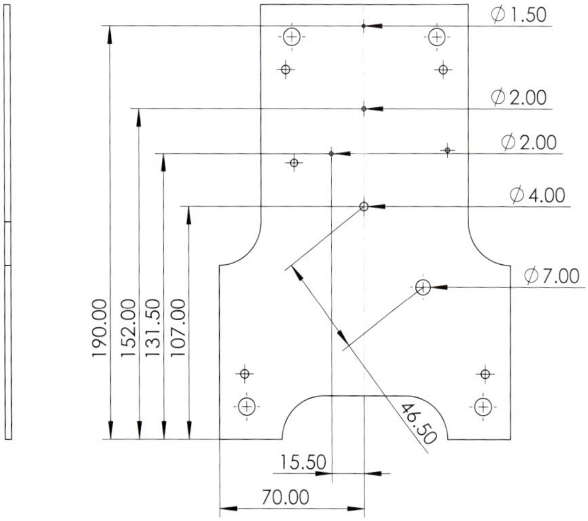

Front plate train for deadbeat escapement. 107mm centre arbor. 152mm escape wheel.

This will be sufficient at this stage. If you have a depthing tool it now needs to come into play. If not, proceed to the paragraph headed 'Setting the train without a depthing tool' on page 122.

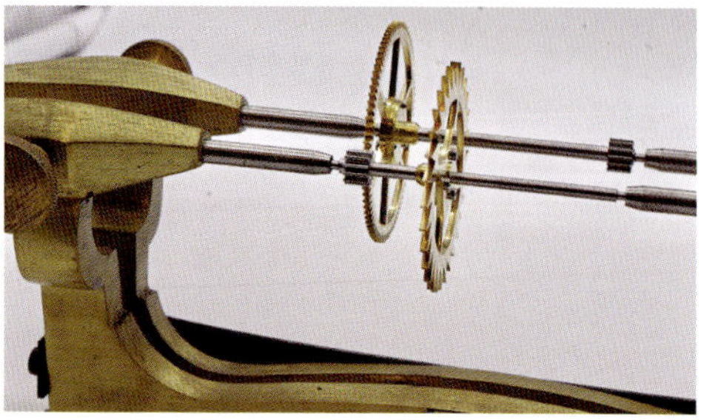

Mount the escape wheel in the depthing tool.

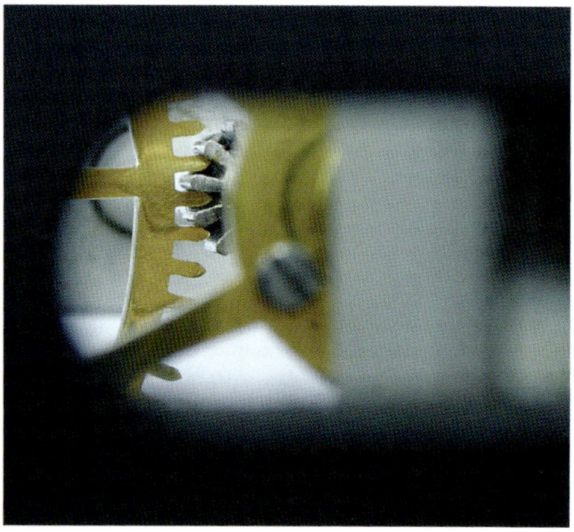

A different depthing tool to the one shown in the previous photo is used here to give a view through the tool to illustrate the correct depth for the centre wheel depthed into the third wheel pinion.

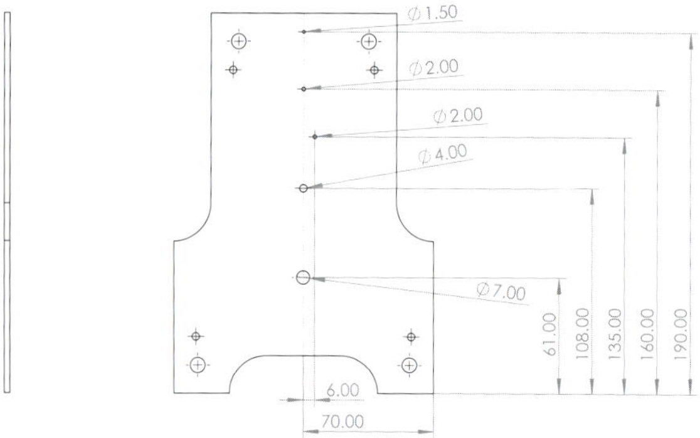

Location of train wheels' recoil escapement. Centre wheel 108mm. Escape wheel 160mm.

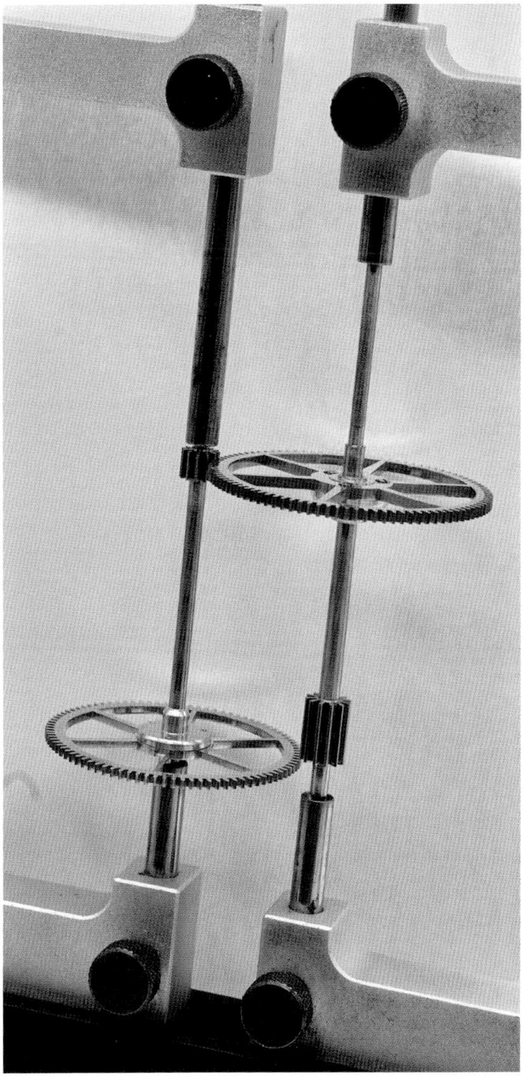

The author has another type of depthing tool to illustrate the correct depth, as it makes a clearer photo.

Using the depthing tool

Take the escape wheel assembly and mount it in the clockmaker's depthing tool.

Take the third wheel assembly and mount it on the other side of the depthing tool as can be seen in the photo.

Now, using the adjusting knob, carefully bring the third wheel teeth into mesh with the escape wheel pinion. As you begin to turn the adjusting knob the two sections of the depthing tool begin to close together and thus the wheel and pinion will begin to engage. As you deepen the mesh with the adjusting knob the mesh will become smoother. If you deepen the mesh further you will find it will begin to feel rough before it finally binds. Try not to reach the point of mesh where it binds, as you may spoil your work or damage the tool.

It is a question of trial and error, adjusting the depthing of the wheels until you reach the point when they are running at their smoothest. Take some trouble over this: it is worth getting the wheels to run as smoothly as possible. It is not easy to see down the length of the spindles of the depthing tool to visually appreciate the depth of the mesh. Using a different design of depthing tool, the illustration on the preceding page gives an indication of the correct depthing for these wheels.

(**Note:** J. Malcolm Wilde's publication referred to earlier in the text has diagrams and notes for depthing clock wheels.)

Setting the wheel depthing on the plate

On the reverse end of each adjustable spindle of the depthing tool there is a sharp point. It is important to keep these ends in good condition as they are used to scribe the plates indicating where to drill the pivot holes.

The depthing tool is used as a pair of dividers to mark an arc on the clock plate. Take care not to move the depthing adjusting screw as you begin to mark the plate; also, remove the wheels when using the tool in this way.

Hold the tool upright, adjusting the length of the spindles to make this so. Place one point in the hole drilled for the escape wheel arbor and score a line in between the centre wheel and escape wheel holes, to the left of the centre line.

Next, place the third wheel with the centre wheel into the depthing tool and adjust the depthing of the centre wheel with the third wheel pinion until the wheels run together at their smoothest point, shown left. Remove the wheels from the tool, use the centre wheel hole to scribe a line in such a way as to cross the previously marked line.

Where the two lines cross is the centre point for the location of the third wheel. Centre-punch and drill this hole as the previous two have been – 0.1mm smaller in diameter than the smallest pivot you have made on the third wheel arbor. (*See* photo overleaf.)

All that remains now is to depth the centre wheel and the great wheel in the depthing tool; mark and drill these holes as above. The location of the great wheel arbor, and therefore the winding square, if you are making the key-wound version, is one of personal preference, depending on where you would like the

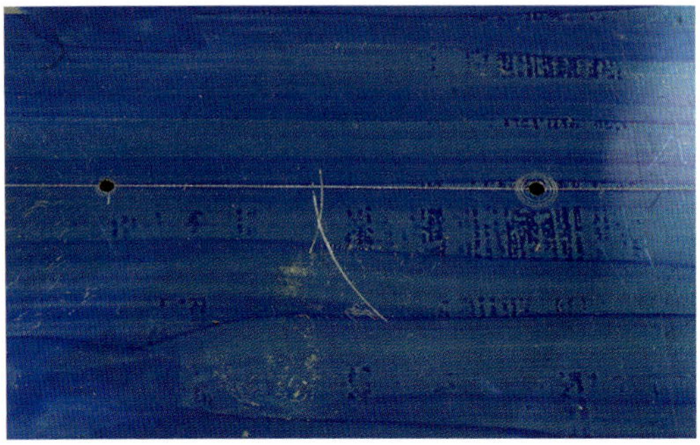

Shown are the two drilled pivot holes – one for the escape wheel and the other for the centre wheel. Using the depthing tool as described produces two crossing lines, this is the correct location for the third wheel pivot hole.

key to pass through the dial of the clock. (It must obviously be correctly depthed to the centre arbor pinion.) The drawings show the barrel arbor to be mounted on the centre line of the movement; however, the movement photographed has the barrel arbor offset to allow the same set of plates to be used for both designs of the movement.

At this point all the arbors are located and drilled through on both plates relative to each other and in regard to the pillar holes, although the holes will be too small for the pivots to pass straight through.

Next the plates are separated and the pivot holes are drilled out individually to within 0.1mm diameter of the pivot diameter fitting the hole. With the larger diameter holes for the barrel arbor, it is best to work through several drills to ensure a good clean hole exactly where you require it. A slow speed is also important in order to allow the brass to clear the cutting edge of the drill. This gives a smoother and cleaner cut – it just requires a little more patience.

Setting the train without a depthing tool

Whilst the depthing tool is very useful indeed, it is also an expensive piece of equipment, which you may choose not to purchase for making just the one clock. In this movement the wheels and pinions have been set near to each end of the arbor in order to make depthing them as easy as possible without the use of the depthing tool.

As with the depthing tool instructions above, the escape and centre wheel positions are known and can be marked and drilled. To depth the third wheel to the escape wheel pinion and the third wheel arbor pinion to the centre wheel, a piece of 3mm-thick brass is required. The same piece of brass can also be used to set the depth of the centre wheel pinion to the great wheel on the barrel.

As a starting point, use the dimensions shown in the plate layout drawings above, drilling two holes whose centres are the distance shown on the drawing – one of the diameter of the front centre pivot, the other the pinion end of the third wheel arbor. Place the respective pivots in these holes and look at the depthing of the wheel teeth into the pinion leaves. Is there clearance between the pinion leaves of the third wheel and the teeth of the centre wheel? Is there clearance between the tip of the centre wheel tooth and the root (bottom) of the third wheel pinion?

Referring back to the photo on page 120, does the depth of the wheel and pinion concerned look as in the photograph – do they look closer, or more distant? Drill a further hole either slightly nearer or further than the starting point and try again. It is a question of trial and error, but worthwhile taking care over as the depthing needs to be correct in the clock movement. Use the same technique for depthing the third wheel with the escape wheel pinion. This will give you the centre distances required for the placing of the third wheel pivot holes. Mark these using a pair of dividers and drill them through 0.1mm diameter less than the smallest pivot on each arbor. The same technique is then used to set the depth of the centre wheel pinion and great wheel.

CUTTING THE OIL SINKS INTO THE PIVOT HOLES

An 'oil sink' is required to form a small reservoir of oil to lubricate the pivot. You can make your own flat countersink cutter, or purchase commercially produced flat countersinks, roller sinks or a multi-bladed circular cutter known as a burr.

Simply place the countersink on to the *outside* of either plate at a pivot hole and apply gentle pressure whilst turning the tool. It will cut a countersink into the surface of the brass as shown on the following page. If using a burr to form the initial countersink, there will still be a small burr at the edge of the hole, which will need to be removed with a flat countersink.

Correcting an oversize pivot hole

If you overdo it, and the pivot hole is either at an angle or too loose, all is not lost as the hole can be bushed.

As a guide for bushing a pivot hole that is too slack, drill the hole in the plate with a drill 2mm wider than the diameter of the pivot being fitted. Then make a brass bush in the lathe (in the same manner as making a wheel collet brass blank) 0.5mm thicker than the clock plate and of an outside diameter of the diameter of the pivot plus 2mm. For example, if the pivot diameter is 1.6mm, the bush will be 1.6 + 2, so 3.6mm in diameter. Whilst the bush is in the lathe, drill a hole 0.1mm under the size of the diameter of the pivot to be fitted, and then fit the bush, which should have an outside diameter to be a tight fit into the hole already drilled into the plate. The extra 0.5mm on the length of the bush is to enable it to be riveted into the clock plate. Use a domed-headed punch on the outer end of the bush to rivet the bush in place. Repeat the operation using the five-sided broach and take smaller cuts, checking the fit of the pivot frequently. You will then need to tidy the rivet end where it is proud above the surface of the plate. This can be done by either filing it or using a countersink to cut it down. If possible, use a countersink, as this can be used on its own and makes a clean finish, forming an oil sink in the end of the bush.

The oil sink cut into the pivot holes.

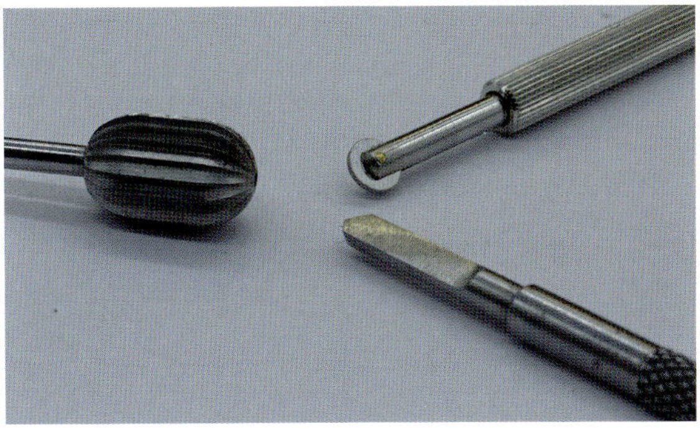

Three different types of countersink: left is known as a 'burr'; top-right is a roller sink, which cuts a radius with a polished finish into the surface of the brass. Lower right is a flat countersink, which cuts a straight-sided counter-sunk hole. The flat countersink is also used to remove the small burr that forms when using a broach to open the pivot hole to the correct diameter.

END SHAKE

The other adjustment that may be required is that of the arbor needing to move freely from the inside of one plate to the other. In other words, the arbor should not be clamped by the plates when they are put together. It is necessary for the arbor to be able to float in between the plates, this is called 'end shake'. Quite simply, if there is no end shake the clock will not work.

The end of the hand mill – sometimes referred to as a star cutter.

Another view of the end mill, showing how the centre 'spike' will locate in the pivot hole.

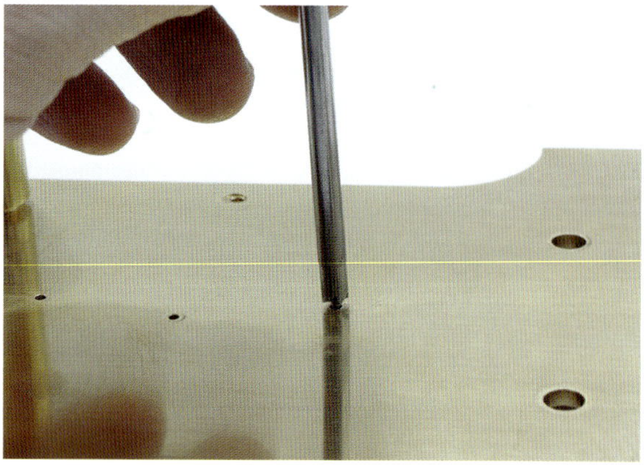

Use the cutter, holding it vertically. It appears to be slanting in the photo; this is camera angle only. Apply a light pressure whilst rotating the cutter clockwise.

The lack of end shake will be caused by the distance from one pivot shoulder to the other being greater than the length from one shoulder to another of the movement pillars. There are two ways to correct this problem. The first is to turn some steel from one end of the arbor at the pivot shoulder point. The other way is to mill a small amount of brass from the inside of one of the clock plates, using a hand-held end mill. This effectively increases the distance between the plates at this point. Milling a small amount from the inside of the plate does mean you avoid the risk of damaging the finished pivot already made and fitted to the hole. Only a small amount may be removed in this manner. If more than 0.5mm is required, the shoulder of the pivot will need to be turned away rather than reduce the thickness of the brass bearing.

Use the end mill as shown. End mills are available either in a clockmaker's staking tool set or can be purchased individually.

Once all the holes are drilled and the pivots fitted to their respective pivot holes, assemble the clock movement made so far and push the wheels round. Everything should run smoothly, and if so, good job done!

If the wheels do not turn smoothly, then first check there is end shake on all the wheels, and secondly that none of the pivot holes are binding – this will become apparent as you check the end shakes.

Having checked these points, if some stiffness still remains it might be that one of the pivots has been bent whilst assembling the movement. Lay the clock movement down and carefully remove the front plate, then each pivot can be checked in turn. If these are all true, place each wheel in the plates and see that it is free. Having checked these, if there is still some friction in the train check each wheel and pinion depthing, to make sure the plate bearing holes were drilled through in the correct location. Having carried out these checks and made appropriate adjustments, the clock will run smoothly.

11

The Motion Work

THE MOTION WORK IS the term given to the twelve-to-one reduction wheel work located on the front of the front plate, controlling the rotation of the hands. The minute hand, mounted on the cannon pinion, is driven from the centre arbor and therefore rotates once per hour. The hour hand is geared via the cannon pinion, reverse minute wheel and pinion to one turn in twelve hours. Therefore the motion work consists of three wheels, one pinion, a tension spring and a bridge – the hour wheel bridge. The wheel work and tension spring are shown in the photo on page 129. The hour wheel bridge is discussed in the chapter on making bridges.

Traditionally, the hour and reverse minute wheels are not crossed. The author tends to cross them as it reduces the weight of the going train just that little bit more. When the motion work is visible their appearance is much more elegant when crossed as well as the practical issue of weight saving. The cannon pinion wheel is left solid in order for the hand tension spring to have a solid band to act upon regardless of how far it is compressed by the minute cap in due course. The subject of the minute cap is covered in a later chapter. The cannon pinion consists of a brass pipe, with a hole drilled centrally to allow the centre arbor to pass through. The pipe is riveted to the cannon pinion wheel, and as an assembly mounted on the centre arbor, it passes through the hour wheel bridge pipe. The minute hand is mounted on the end of the cannon pinion assembly, usually on a square seat to stop it slipping from position.

The hour wheel is mounted – riveted – on a brass tube, the front section of which is the mounting point for the hour hand. It is quite a straightforward turning exercise, as can be seen from the drawing.

The hole in the hour wheel tube is the running surface for the hour wheel to rotate on the hour wheel bridge. It is important that this hole is smooth and true. When drilling, use a drill 0.2mm (8.8mm in this instance) in diameter smaller than the diameter of the hour wheel bridge and use an engineer's reamer to cut the final diameter of the centre hole. If the reamer is not available, drill the final full diameter, taking your time to ensure the smoothest cut.

The cannon pinion is the arrangement by which the drive is transferred from the centre arbor to the motion work. It is also the point at which a 'clutch'-type mechanism is fitted to enable the hands to be set to time. There are two parts to the cannon pinion: a brass tube, onto the end of which the minute hand will be fitted; and a brass wheel.

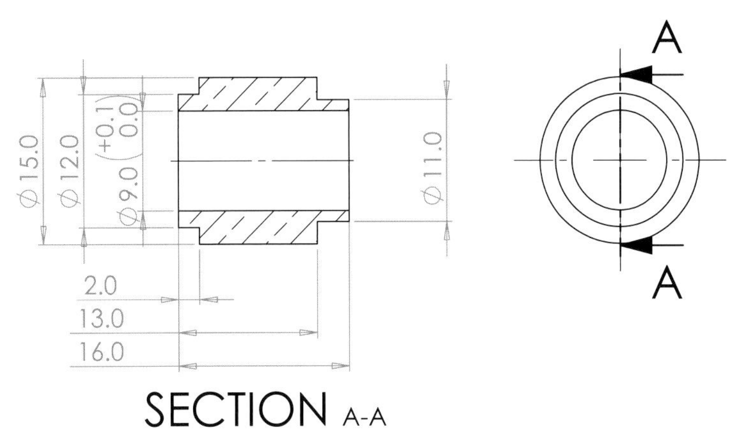

Hour-wheel pipe.

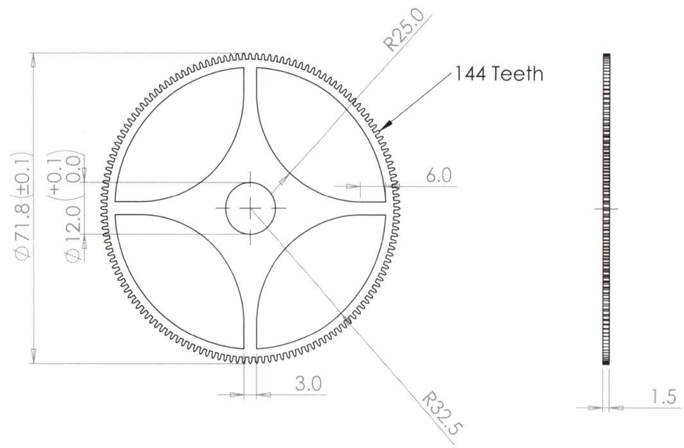

Hour wheel 14 crossings.

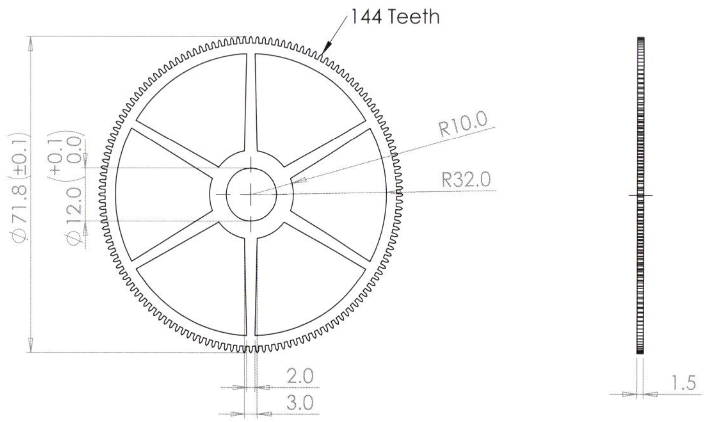

Hour wheel 6 crossings.

The cannon pinion fitted to the centre arbor.

The reverse minute wheel ready to be fitted to the pinion.

The wheel has the same dimensions as the reverse minute wheel and so the two are cut together. The two wheels are 2mm in thickness so that they stay in mesh when the tension of the spring beneath the cannon pinion is compressed to secure the hands. In compressing the spring the cannon pinion changes its height compared to the reverse minute wheel. The wheel thickness allows for this.

As the wheel is cut and remains solid, it is effectively made, other than smoothing and polishing the sides of the wheel.

The tube section is a little more involved as it requires turning, drilling and a square forming on the end to accommodate the minute hand. The dimensions are shown on the drawing. Begin with a piece of brass bar stock of at least 9mm in diameter.

Mount the bar in the lathe in such a way as there will be 34mm of bar clear from the collet or chuck. Drill a hole of diameter 3.1mm through the centre of the bar, and turn the outside profile as shown in the drawing.

This gives the seat for the cannon pinion wheel to fit onto, and also the pipe with clearance to pass through the hour wheel bridge. Finish the surface so it is smooth; there is no need to polish it. While still mounted in the lathe, turn a shallow groove 2mm in from the opposite end to where the wheel will be mounted. This is a line guide for filing a square on the end of the tube to mount the minute hand.

The remaining task is to form the square for mounting the minute hand. The author prefers to part the cannon pinion from the bar stock and hold it in the vice to file the square. It could equally be left mounted on the bar and filed whilst being held in the lathe.

Use the guide line to aid the filing process; try not to file over this line. Care is required: using the square needle file is a good choice of tool, as it will allow you to see exactly what you are doing. Take your time, file three or four strokes and then turn the cannon pinion 90 degrees and file again. Work your way until all four sides have been filed. Continue this process, using the vernier gauge to measure across the flats often, as was done when filing the winding square, until the four corners being formed meet. This should result in an even and relatively central square.

Having made the two sections they are now riveted together using the split brass stake previously made. It could also be assembled using Loctite 638. The completed cannon pinion should slide down the extended centre arbor, without binding, but not be too loose.

Making the reverse minute wheel and pinion assembly can be done by either making the pinion as a short arbor – as the clock wheels are, and using the front plate and a brass potence (a one-sided bridge) as the pivot points – or by mounting them on a steel post. The post method is quicker and more straightforward to mount. Whilst the author has used both types in clocks produced by the workshop, described here is making and fitting a steel post.

Begin by turning a piece of EN8 steel to an outside diameter of 7mm, and then turn to a diameter of 2.3mm for a distance of 15.5mm (shown overleaf).

Polish this to give a finished diameter of 2.2mm. When making the post it is important the surface of the post is well finished and care is taken when drilling

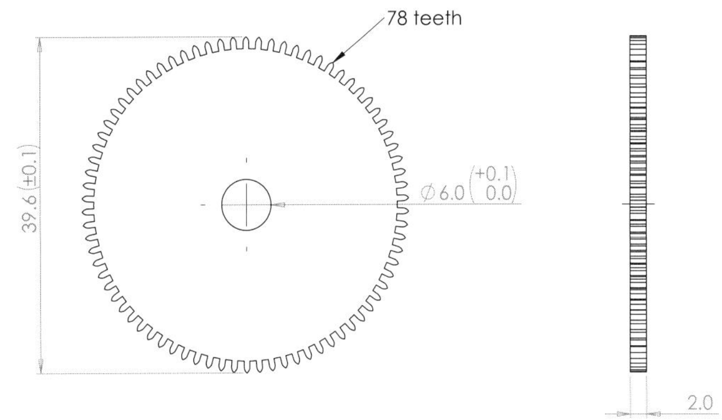

Cannon pinion and reverse minute wheel.

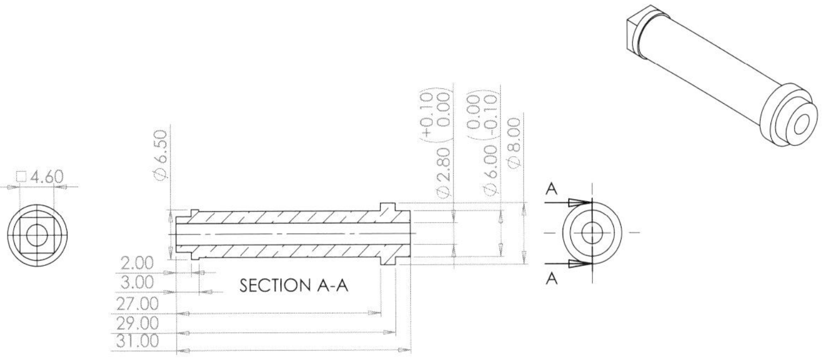

Cannon pinion tube.

The pillar turned.

The pillar with the square filed mounted in the vice.

the pin hole in due course. If you wish you could thread the end of the post using an 8BA die, and fit a small nut rather than drill a pin hole.

Reverse the workpiece in the collet or chuck, and turn the diameter to 3.6mm for the length indicated. Using the 4BA die, form a thread on this section of the post.

The section turned 7mm in diameter is for forming a square by filing in the vice, using brass soft jaws, as the other squares have been filed – use a fine-cut hand file. The square is used as a purchase for screwing the pillar in and out of the plate.

Reverse minute pillars are also available commercially. They would need adjustment, however.

If you are using a pin to secure the pinion, the hole is drilled through the end of the post ready for a pin to pass through to hold the reverse minute wheel and pinion in place. The diameter of the hole is 0.7mm, making the 8BA threaded option rather more straightforward to produce, and the author's preference. As with the other holes drilled into round stock, you can make a guide to help. When turning the diameter of the post, turn the front section 20mm long and cut off 4.5mm of this, leaving the front of the post 15.5mm long as shown. Drill a hole in the centre of the cut section diameter 0.7mm – the diameter of the pin hole. Set the post and guide in the machine vice as done for drilling the mounting holes for the bottom movement pillars, and drill through the post using the 0.7mm diameter drill. Do not over-clamp the pillar in the vice, as this will cause damage to the smooth surface previously prepared for the rotation of the reverse minute wheel pinion.

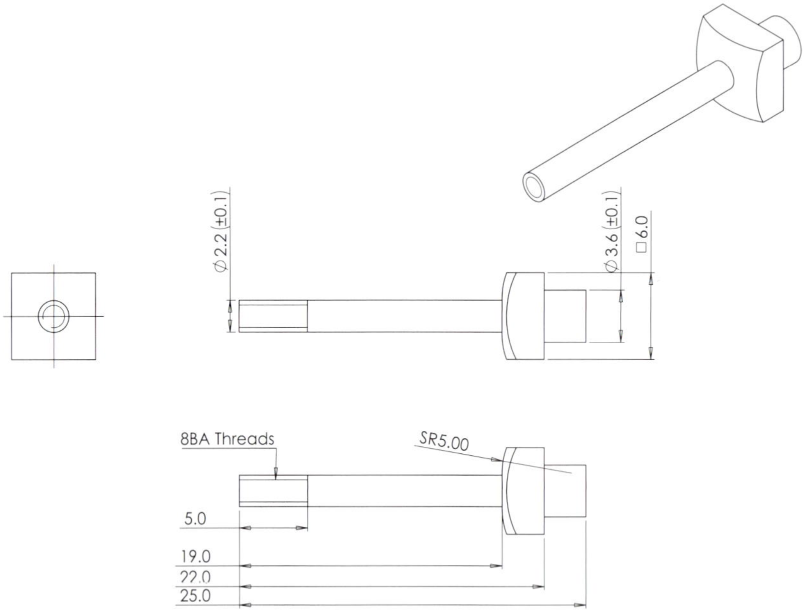

Minute-wheel post.

The last part of the motion wheel work is to make the tension spring that will act as the clutch for moving the hands. The hands need to hold in position to be driven by the clock, but you need to be able to turn them to set the clock to time. The amount of tension is controlled by the spring we are discussing here.

The drawing below shows the design and dimensions of the spring. Take a piece of 1mm-thick brass sheet and hammer it to form a spring in the same way as the click spring brass was treated. Mark onto the brass the shape indicated, centre-punch the centre of the hole and drill it to a diameter of 3mm – to match the diameter of the front section of the centre arbor.

Cut the shape out and tidy the edges. Place the cannon pinion on top of the spring and push it against the spring to see if the spring will give – which it probably will not at this stage. Reduce the thickness of the two sides of the spring until you can push the cannon pinion onto the spring and feel it 'springing'. It is important that the spring does not collapse, but pushes the cannon pinion back as you release the pressure. Once the spring is working well it can be finished off as you wish. The important areas to make smooth are the ends of either side of the spring where it comes into contact with the underside of the cannon pinion wheel. The spring is not visible in the finished movement, so finishing the remaining surfaces of the spring is a personal choice.

In due course the hour wheel bridge will be fitted, allowing the hour wheel to be mounted. See page 166 for a photo of the finished motion work mounted on the chain-wound version of the movement.

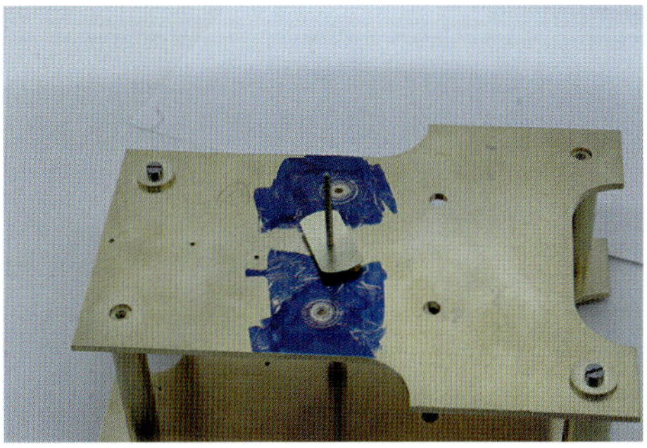

The finished spring fitted to the centre arbor.

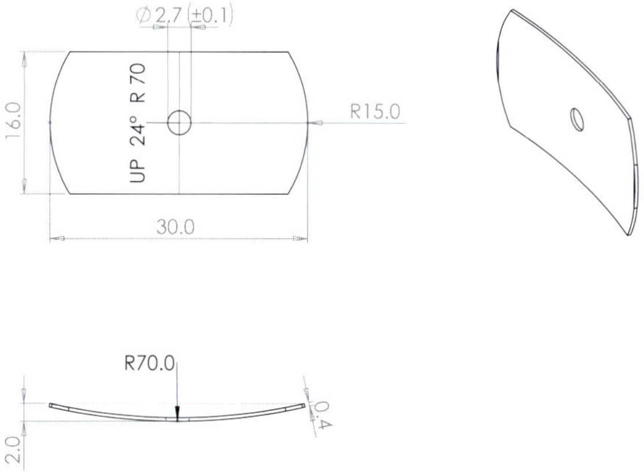

Cannon pinion tension spring.

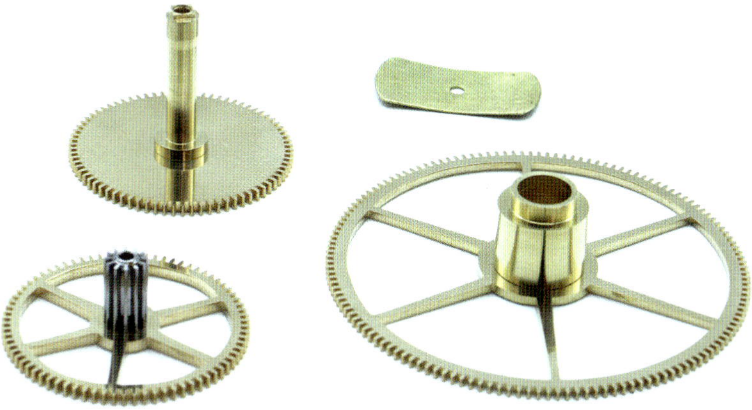

The finished parts of the motion work. Starting at the top left-hand corner and moving clockwise, they are: the cannon pinion; hand tension spring; hour wheel complete on its tube; and finally, bottom-left, the reverse minute wheel crossed and mounted on its pinion.

12 Making the Various Bridges

TO CONTINUE THE MOVEMENT we need to construct two bridges to help locate the various parts of the movement. These are the pallet and hour wheel bridges. The dimensions for the bridges to fit are important, but the actual style of each is another opportunity for you to make the movement unique to you. Each of the bridges shown in the photo have been fabricated by silver soldering the sections together. If you have a milling machine, both these bridges can be cut from solid brass square sections, only leaving the hour wheel bridge pipe to be turned and fitted. The photos show both types of construction.

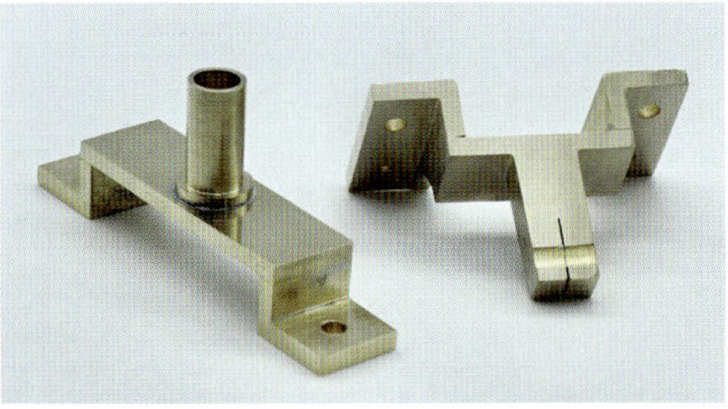

The hour wheel and pallet bridges constructed by fabrication.

The pallet bridge milled from a solid section of brass.

The hour-wheel bridge milled from a solid section of brass.

THE PALLET BRIDGE

The job of the pallet bridge in a typical clock is twofold: to hold the pallets in position for the action of the escapement to be correct, and also the point from which the pendulum is suspended. It therefore needs to be rigid and securely mounted to the movement back plate. To aid the rigidity of the mounting it is usual to have large 'feet' on the bridge. In precision timekeeping clocks the pendulum is suspended from a very sturdy bracket mounted on the inside of the back of the clock case. This does require the back of the clock case to be very strong in construction, and preferably attached to a solid wall. For this project the pendulum is mounted on the rear of the pallet bridge. This has the advantage of allowing the movement to be run on a test horse or wall bracket without the need to mount the pendulum separately.

Fabricating the bridge

In the absence of a milling machine, bridges can be fabricated by cutting and silver-soldering sections together, or by sawing and filing the shape from a solid piece of brass. Due to the size of the pallet bridge 'feet' it is better fabricated, as cutting from solid would be quite a job. The process for fabrication is described below.

Use brass already formed into 90-degree angle material, as it is a much more straightforward and less time-consuming process to fabricate the bridge. It will be necessary to have a thickness of 3mm for the brass to be strong enough and allow some removal of brass when finally shaping the bridge.

Begin by marking-out the rear section of the bridge as shown in the photo. Next the two side sections which will form the feet are marked-out. Use the steel rule, marking blue, scribe and set square to mark out the sections to be cut.

Neatly saw the rear section and the two sections of angle, these will be the feet that will attach the bridge to the plate. Do not finish any of the cut surfaces at this stage. *See* the drawings for dimensions.

The angle marked-out ready to cut the central section.

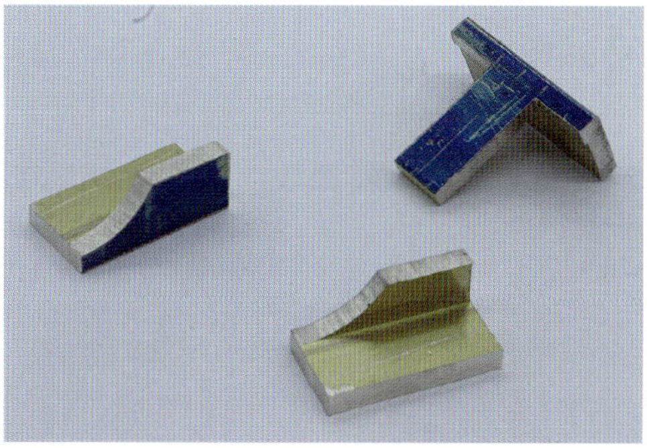

The three sections are cut ready to silver solder.

Silver soldering

Important note: DO NOT use oxygen–propane bottled gas as used here without flash-back arrestors fitted.

Once the sections have been cut to the shape shown on the drawing, the next task is to silver-solder them together. The process of silver soldering is straightforward, but it is imperative that the brass is clean and that the correct temperature can be reached. An oxy-propane torch gives a hot clean flame. Care should be taken when lighting such a torch, as propane takes a few seconds to settle down to burn due to the gas being in liquid form in the cylinder and needing to change 'state' to a gas as it exits the cylinder. A butane blowtorch could be used.

It is also important to use the right flux. The author uses borax, purchased in powder form and mixed with cold water to a paste for application. This must be mixed in a clean container each time you need it. It does not keep, so mix what is required to avoid wastage. Do not be tempted to re-use it; you will only have a burnt and sticky mess to

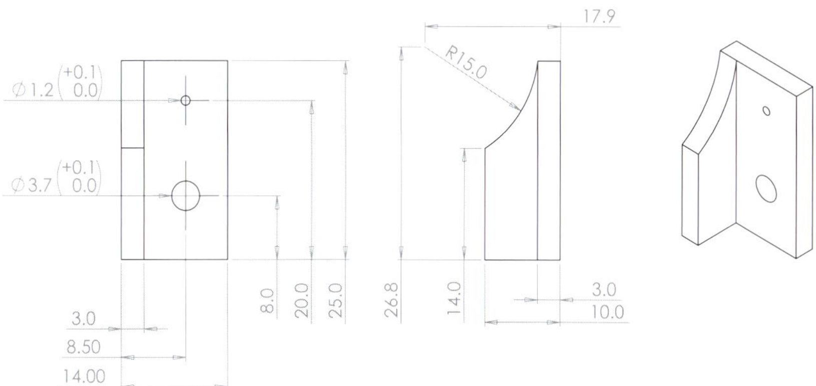

Pallet bridge LH foot.

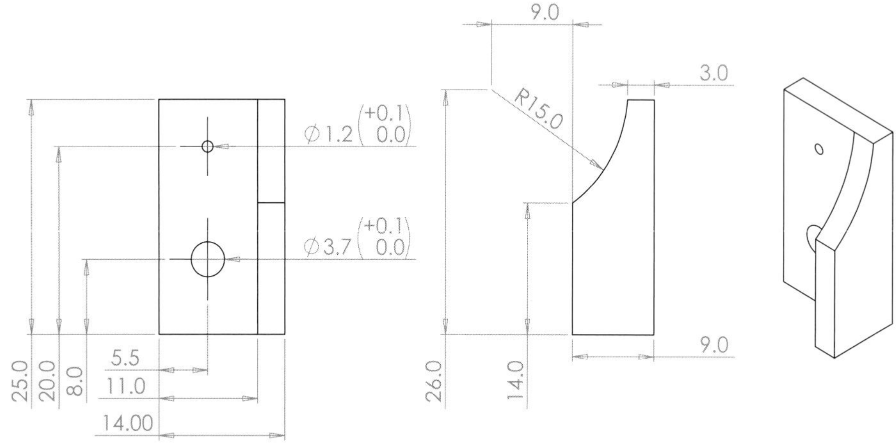

Pallet bridge RH foot.

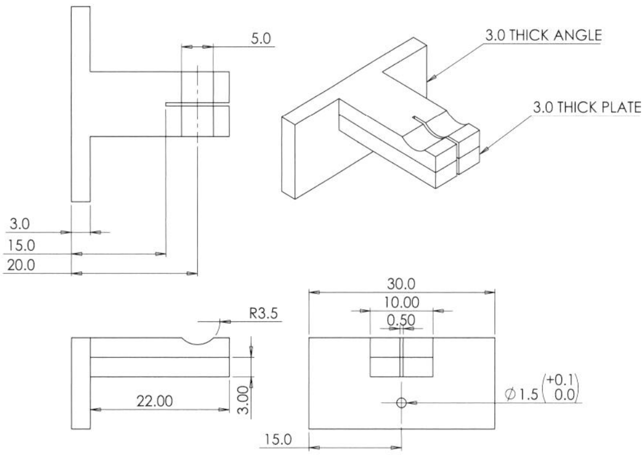

Central section pallet bridge.

clean up before beginning again! A hearth is required for heating objects in a safe environment, and is straightforward to arrange. Normal fire bricks or matting such as used with wood-burning stoves can be used to line the hearth. Do not use bricks from an old storage heater though, as these are designed to absorb heat rather than reflect it.

Clean all the surfaces to be joined so that they are bright, clean metal at the point where they are to be joined, and carefully lay the sections together. A fine file is a good way to clean ready for silver soldering. It is important to arrange them in such a way so that they do not move when heated. A good tip is to drill the parts and pin them together with brass pins so they stay exactly in position when the heat, flux and silver solder are applied.

When the assembly is heated the flux will boil – moving any parts that are not secure out of alignment – which can be very irritating! Brass pins will not show on the finished bridge.

Having drilled and pinned the sections together, coat the surfaces to be silver soldered with a generous layer of freshly mixed flux. Place the assembled parts on the hearth and heat the brass until it is red hot – do not overheat it as it will melt or burn. Apply the silver solder rod, the heat from the brass will melt it. Do not heat the silver solder directly with the torch as it will melt and probably form a blob on the workpiece – this is unhelpful and may result in having to start again if it will not melt into position as heat is applied.

Let the workpiece cool down before beginning to clean the work. It can be finally cooled in cold water. As can be seen from the illustration on page 133, parts that have been silver soldered together have a residue of sticky and also rather glass-like flux, which needs to be removed. This is simply a matter of running boiling water over the parts and the flux will dissolve.

Shaping the pallet bridge

The next task is to file and cut the bridge to the shape shown. Care is required when shaping the bridge to ensure a sharp clean and straight edge to the workpiece. Begin by cutting the rear section of the bridge with the piercing saw to form the pendulum hanging bracket. Once this is done the remainder of the shape is formed by filing. File just a few strokes, and then check the workpiece. Checking often will save going too wrong if the angle of your filing is not straight. Using soft jaws in the vice will help not to add marks to your work. Once shaped, use the piercing saw to cut a slot to allow the pendulum suspension spring to fit.

Next, drill the holes at the location shown on the drawing. Then using a tapered steel clock pin as a guide, use a five-sided broach to ream the hole until the pin fits neatly

1

The pallet bridge parts laid out ready to silver solder. The flux can be seen as white paste.

2

The bridge once heat has been applied and the solder has flowed.

3

The bridge – washed with boiling water to remove the flux residue.

two-thirds of its length into the hole. Take care to hold the broach straight, as the pin needs to pass straight through into the rear of the back movement plate in due course.

Having shaped the bridge it is necessary to apply a finish. You may choose to polish all the surfaces, or perhaps to polish the facing surface and put a grained type of finish on the edges. The author prefers to have a grained finish on the facing surfaces and to polish the edges. This gives a smart finish and reduces the risk of rounding the corners of the workpiece when polishing.

Note the point where the pendulum will locate is now curved in shape, rather than as an angled slot.

THE HOUR WHEEL BRIDGE

The process for fabricating the hour wheel bridge is similar to that of the pallet bridge, the drawings show the dimensions for each section to be made and how they are to be silver-soldered together.

It would be a practical option to cut or mill the hour wheel bridge from a solid section of brass, as it does not require such a large 'foot' as the pallet bridge. When cutting from solid, mark-out the solid section and begin by sawing down each section with either a hacksaw, or as the author would do, with a piercing saw. It will take a little time and care to use the piercing saw, but the result will require much less finishing, and due to the cutting action of a piercing saw you have more control over it.

The drawing gives dimensions and also the location for the securing screws and steady pins to pass through.

Having produced the basic hour wheel bridge by whichever method chosen, it requires the fitting of a brass pipe to carry the hour wheel assembly. This is a straightforward turning job from brass bar to the dimensions shown. Note the pipe must not be secured into place yet, as the hole will require drilling to a larger diameter in due course.

For the moment the hole in the centre of the bridge is drilled to the diameter of the extended front section of the centre arbor – 3mm. This will help to locate the hour wheel bridge centrally with the centre arbor.

With the centre arbor placed between the plates, screw the plates and pillars together; place the hour wheel bridge over the centre arbor. Provided it is a snug fit, use a drill of 1.2mm held in a pin vice as a marker through the steady-pin holes. Dismantle the

The pallet bridge being roughly filed to shape – there is a bit left to do.

Having cut the slot for the suspension spring using a fine blade in the piercing saw, the author is using a square file to file a groove with 45-degree slopes. This will in due course be turned into a curve to match the suspension-spring mounting block.

The slot filed, ready to be shaped to accept the brass head of the suspension spring.

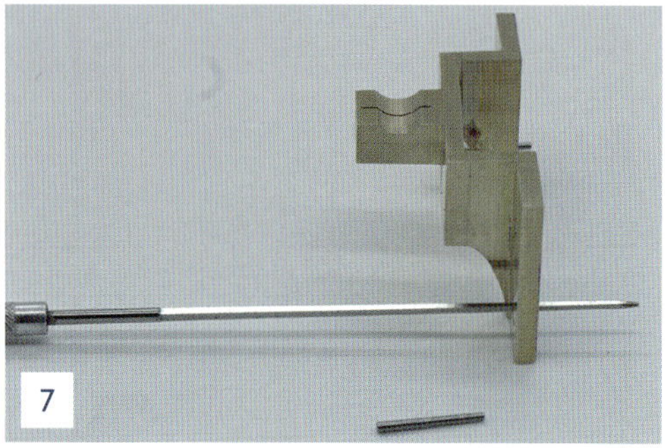

The bridge nearing completion, showing the process for fitting the steady pins.

The hour-wheel bridge – silver-soldered together.

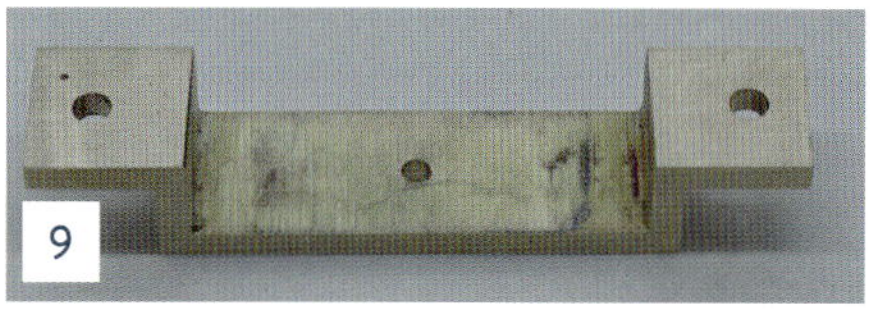

The bridge after the initial filing to shape and the three main holes and two steady-pin holes drilled. The main holes are the two that will secure the bridge to the front plate, and the central one – drilled to the diameter of the centre arbor at this stage. The steady-pin holes help with the initial mounting of the bridge.

The hole in the pallet bridge centre drilled only to the diameter of the centre arbor. This allows the centre arbor to be used to correctly locate the hour wheel bridge centrally to the centre arbor.

movement, centre-punch these holes, drill them and add steady pins to the hour wheel bridge. Reassemble the movement and fit the hour wheel bridge to double-check the steady pins are located correctly. With the hour wheel bridge held in position with the steady pins, mark the securing screw centre locations. Use the drill again as a centre-marking tool, this time 3.7mm in diameter to mark the centre point. Do not forget to use the tapping drill of 3mm diameter to drill the holes marked, as they need to be tapped 4BA.

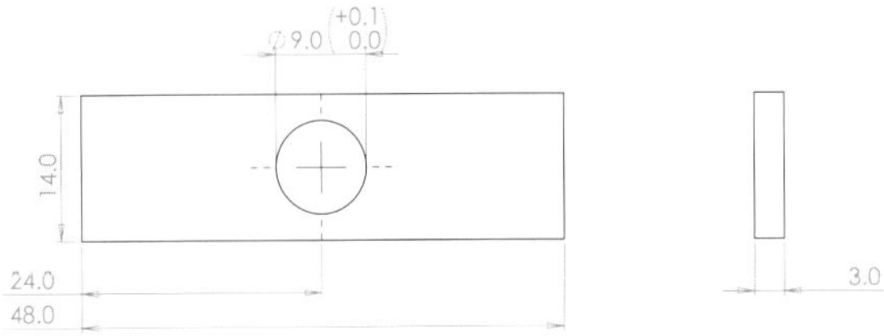

Hour-wheel bridge top. Note: do not drill the hole in the hour-wheel bridge top until instructed to in the text.

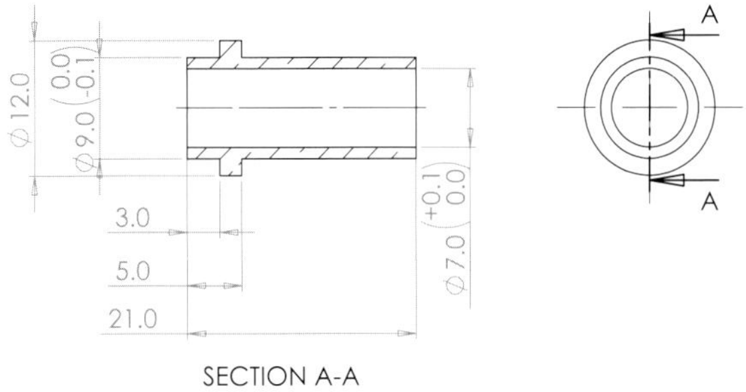

Hour-wheel bridge pipe.

Next, the mid-section of the hour wheel bridge is drilled to locate the turned pipe section. Drill to 9mm diameter, using a 4, 6, 8 and finally 9mm drill. The pipe should be a snug fit and will in due course be riveted, secured with Loctite 638, or soft-soldered into place. For riveting it will be necessary to have a stake with the correct size hole to accommodate the pipe up to the point of the step.

For soldering into place flux is applied to the joint and the bridge heated sufficiently for the solder to melt when it is touched onto the workpiece. It is then necessary to cure the joint to stop the action of the flux, as this causes corrosion if left unchecked. The use of Loctite 638 here is a good alternative.

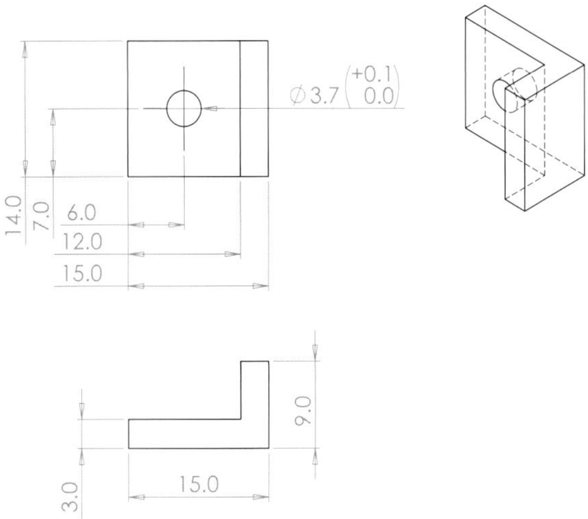

Hour-wheel bridge feet.

CONSTRUCTION NOTES IF MILLING THE BRIDGES FROM SOLID

The author has assumed that the reader who owns a milling machine will be familiar with its use, and has therefore not gone into detail as to how to use a milling machine.

The drawings show the dimensions required. Note the mounting screws are set into the surface of the bridge.

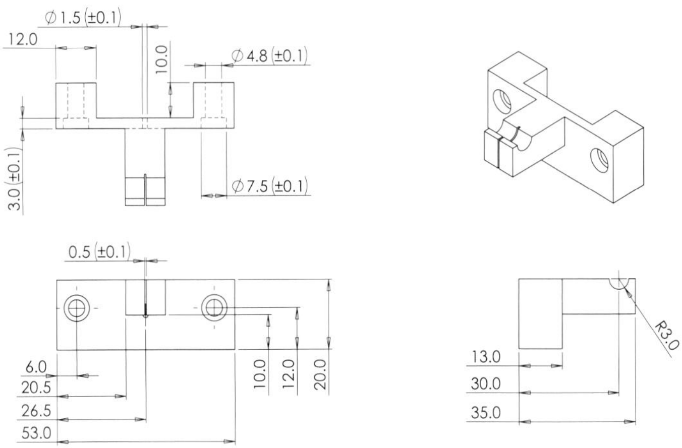

Machined pallet bridge.

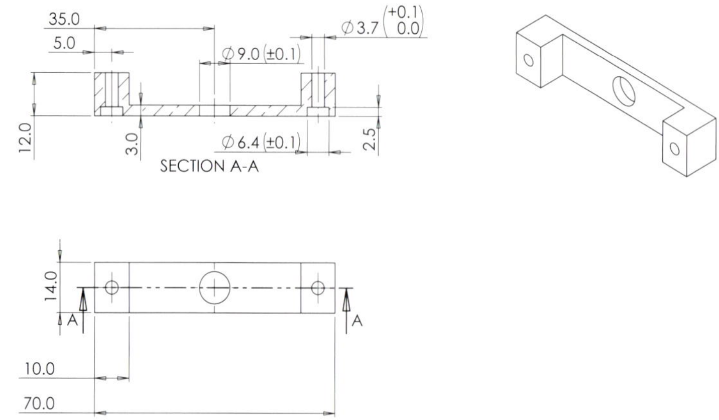

Hour-wheel bridge.

The hour wheel bridge with the central hole diameter enlarged ready to fit the pipe that will carry the hour wheel. Right: the pallet bridge roughly milled to shape. The finished milled bridges are shown in photographs at the beginning of the chapter.

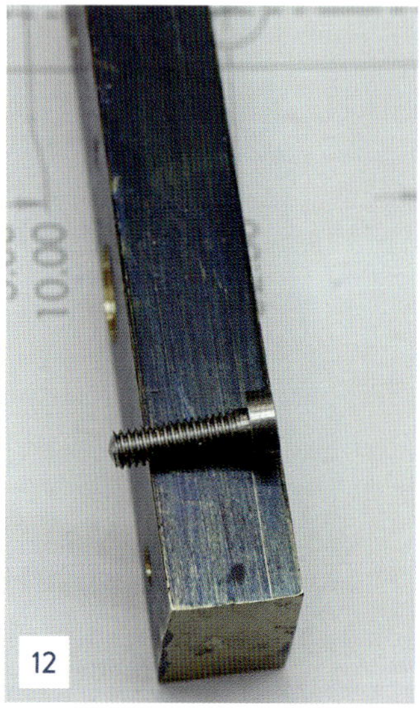

The bridge securing screw laid on top of the bridge to illustrate how the screw fits into the bridge.

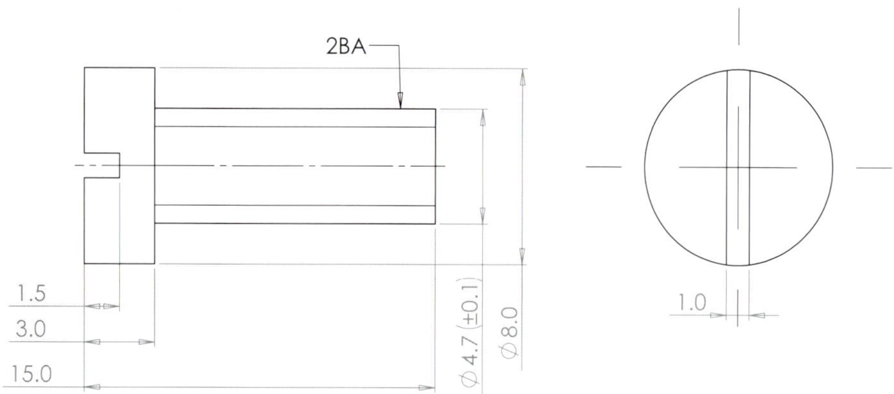

Bridge screw.

Making the Pallet Arbor and Crutch

13

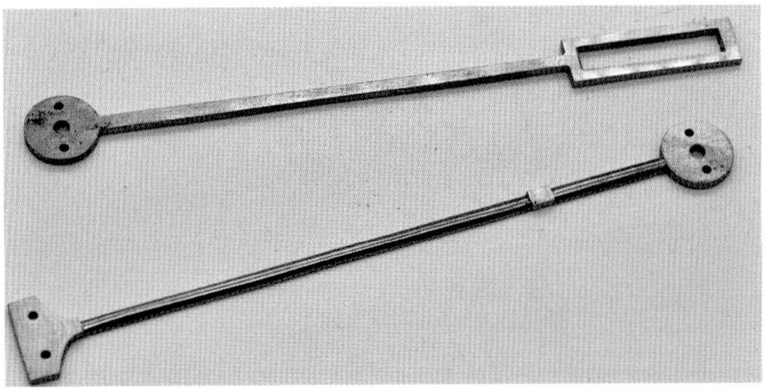

Both types of crutch as described in this chapter.

THE PALLET ARBOR

Before constructing the pallets, the pallet arbor, brass mounting collet and the crutch with its mounting collet need to be constructed. This will allow the placement of the pallets into the movement for final fitting before they are finished.

The arbor is a straightforward turning exercise on the lathe, turning it to the dimensions shown. Take your time turning the pivots, as the better the finish from turning, the less filing and polishing will be required. Bear in mind, the arbor has to be long enough to span the distance between the front plate and the pallet bridge. With this in mind, assemble the movement plates and fit the pallet bridge. The distance between the inside of the bridge and inside of the front plate is the distance required for the pallet arbor between the shoulders of the pivots. To this measurement add a further 6mm for the pivots.

The rear section of the arbor needs to allow the mounting of the crutch. In both variations of the crutch discussed, it is mounted against a brass collet, and held in place with two 10BA countersunk screws. This allows for the crutch to be removed for pivot polishing in the future. It is necessary to choose the design of crutch you wish to make at this stage and then ensure you follow the correct pendulum top section design later when constructing the pendulum. Both the traditional crutch and a version including a beat adjuster are described.

THE CRUTCH

The escape wheel passes impulse to the pallet impulse surfaces and this in turn is passed to the pendulum. This is done via an arm attached to the rear end of the pallet arbor and is known as the crutch. It is also the point at which the beat – tick – of the clock is adjusted to sound even, that is, in beat. When listening to the clock running we listen for the clock to go tick-tock as opposed to tick... tock-tick... and so on. When the beat is uneven, it is said to be out of beat and so a means of adjusting it is required.

The Traditional Crutch

In the traditional longcase clock the crutch is steel, thin and malleable in order to be slightly bendable to adjust the beat when required. It is made in one piece, which is an interesting exercise in cutting, lightly hammering and filing. If choosing this approach, mark-out a piece of steel 2mm thick to the dimensions shown on the drawing.

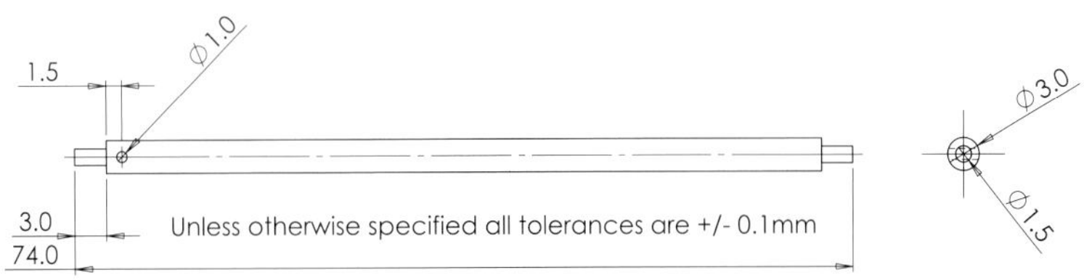

Pallet arbor.

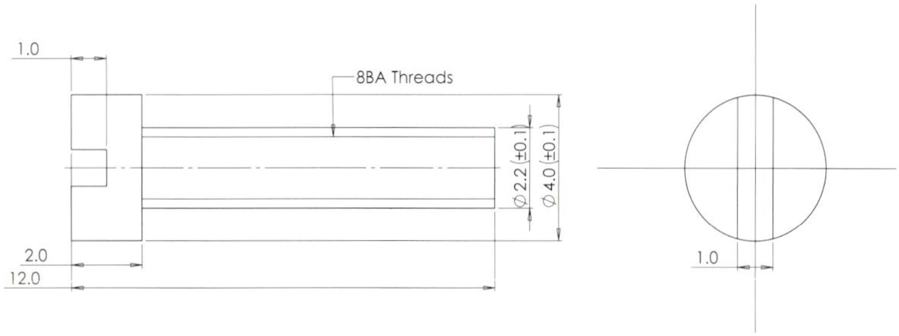

Beat screw.

Begin the construction by centre-punching and drilling all the holes, including a 1.2mm hole in one corner of the rectangular box. This box forms a slot and is the point at which the pendulum block passes through the crutch, and also therefore the point where the impulse is given to the pendulum. The final inside edge of the slot will need to be straight and well finished.

Cut the slot out first, using the piercing saw lubricated with a thin oil. Remember not to try to force the saw to cut more quickly than the blade will allow. Take care sawing into the corners.

Having cut the impulse slot, carry on and cut round the whole shape, taking care to follow the marked lines closely, as it saves time when filing.

The next operation is to make the 2mm square rod section circular. This is begun by holding the crutch in a hand vice and hammering the corners as evenly as you can on the anvil. Once the corners are hammered you have a rather hammered-looking, uneven eight-sided crutch arm. Continue hammering to begin to smooth down the eight corners. Hammering the crutch will work-harden it and also take away the need for quite so much filing to shape it.

Both types of crutch marked-out onto the steel ready to cut.

Rest the crutch arm on the wooden block used to file pins and polish pivots, and begin filing it by rotating it using the technique practised for making steel pins. Take care to have the safe edge of the file against the crutch block in order not to damage it with the file. It will not be possible to file it all the way along by only holding the block end of the crutch. Alternate which end of the crutch is held in the hand vice so that the whole length of the arm is worked. Ideally – but do

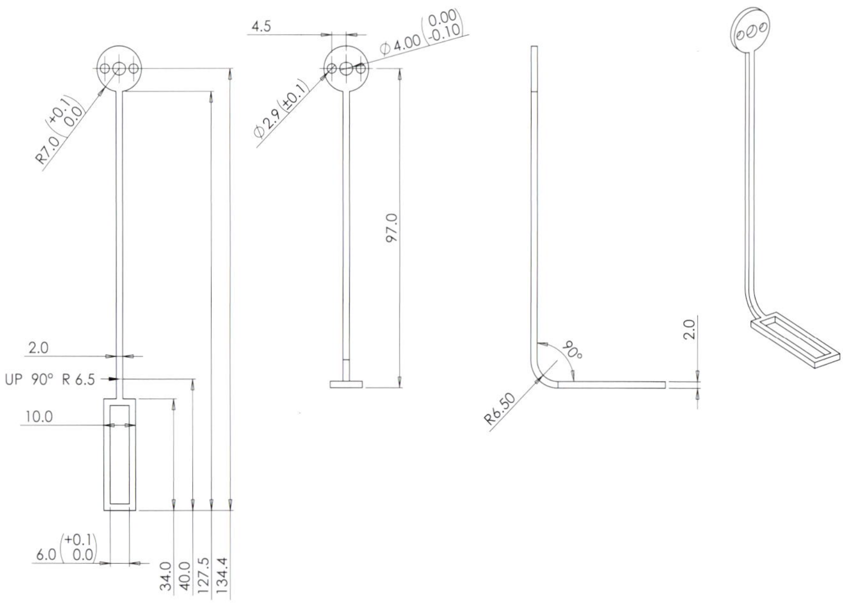

Traditional crutch.

not worry if not – the top end (the end that joins to the pallet arbor) will be slightly thicker than the bottom, so the arm has a slight taper. Remember, the finished rod may need to be bent to put the clock in beat. In due course change to a finer cutting needle file and continue. Once you have a smooth, more or less round crutch arm, change your filing technique and draw-file to finally smooth it before using a buff stick to tidy and finish the rounded section.

Having achieved this, using a fine file, file the inside edge of the rectangular slot section where the pendulum impulse block will be located. Follow the marking lines carefully here as the slot needs to be straight and the same width for its *whole* length. A swinging pendulum swings from side to side, but must also be free to move forward and back. If the pendulum is not free to move in the crutch slot, the clock will stop. The inside width must therefore not have a reduced clearance. File the outside of the block to shape.

The top section of the crutch is the last part to be shaped. This is filed to a circle, which is not as easy as it sounds; especially as it is mounted against a turned brass collet, any discrepancy in the filing tends to show. A real help here is to use a file button.

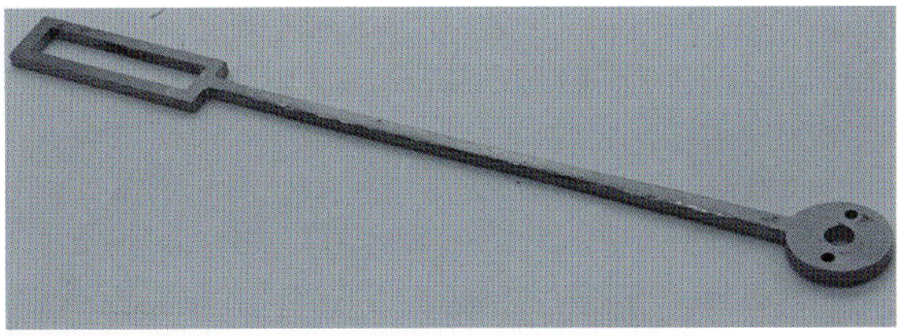

The traditional crutch cut and the corners of the long section hammered (so is effectively eight-sided at this point) – not too closely to either shaped end of the crutch – ready to begin shaping.

The hammered crutch placed on the wooden block used for filing pins and pivots in the vice.

MAKING AND USING A FILE BUTTON

This is a tool easily made in the workshop and is well worth taking half an hour to make. The author has made a number of these over the years. Make it from a steel that can be hardened and tempered. It consists of two parts – one is a bit like a top hat, and the other a washer that fits over the 'hat' section. The dimensions are straightforward. Make the largest diameter of the hat shape and the washer the same as the outer diameter you are filing. Once constructed, check the washer will slide into place correctly and then harden both parts.

For the crutch the hat section of the filing button arrangement is 3mm in diameter and the outer diameter 14mm. To use it, pass the 3mm-diameter section through the hole in the top of the crutch. Next, place the washer on the hat section so the crutch is between the washer and the hat section, and clamp it all in the vice. You can now file down to the washer. As the button assembly has been hardened you will feel the file slip when it tries to cut it. Take care with the safe edge of the file: ensure it is towards the extending rod section of the crutch as you work round the circle or you may catch it with the file, which could result in it being too thin to use.

Another approach to making the crutch is to make the round top section and bottom slot sections and then silver-solder a 2mm diameter steel rod to each end. This saves quite a bit of sawing and filing, but does require some cleaning and tidying following the silver soldering. The disadvantage of this method is the rod section will be relatively soft where it has been heated to red hot to form the silver-soldered joints.

The file button arrangement shown being used to shape the round end of the maintaining power lever.

A complete steel crutch can be bought from a horological supplier's catalogue. Although it does require shaping and altering to fit the pallet arbor, it does save quite a bit of work.

MOUNTING THE CRUTCH TO THE COLLET

Once the crutch is finished to your satisfaction it is time to consider how to mount it to the collet on the pallet arbor. There are already two holes of 1.4mm diameter drilled into the top circular section of the crutch. Place the pallet arbor in the machine vice – using the split stake onto the drill table, with the collet uppermost.

Place the crutch onto the pallet arbor in contact with the collet already mounted. Now, using the 1.4mm diameter bit, drill through one of the holes drilled in the crutch, drilling through the collet (*see* photo on page 90).

Remove both from the vice and tap the collet hole with the 10BA tap to form the thread. Next drill just the one hole you have used in the crutch out to 1.7mm to give clearance for the 10BA screw. You should now be able to screw the crutch to the collet on the pallet arbor with one 10BA screw. Placing the pallet arbor back in the vice and using the crutch as a guide, drill the remaining hole 1.4mm, remove the collet, tap the hole 10BA and drill the remaining hole in the adjuster 1.7mm to give clearance. You will now have two screw holes that exactly line up to secure the crutch to the collet. Countersink these two holes so that a countersunk 10BA screw sits completely flush with the surface of the crutch.

The pallet arbor secured in the split stake ready for drilling.

The finished traditional crutch mounted to the pallet arbor.

The mechanical beat adjuster as described in the text.

CRUTCHES WITH MECHANICAL BEAT ADJUSTERS

In finer-quality clockwork, especially when the dead-beat escapement is used, a form of mechanical beat adjuster is made, rather than relying on bending the crutch.

One way of adjusting the beat is to be able to change the location of the end of the crutch by making it a steel pin, rather than a slot, and mounting it in such a way as it can be rotated on the end of the crutch arm. The pin passes through a slot in the pendulum. This design was used particularly with dial clocks in the early part of the twentieth century.

The object of the exercise was to make a revolving crutch pin held in an adjustable but firm manner by friction. The impulse is given to the pendulum rod through a slot cut into it for the impulse pin to pass through. The drawback of this arrangement, especially with a longer pendulum, is the possibility of the pin moving, causing the clock to no longer be in beat. It is simple to construct, however.

An alternative crutch incorporating the beat adjuster

The type of adjustment preferred and described by the author is a little more involved to make, but gives a positive lock to the beat once adjusted.

The impulse to the pendulum can be given by an impulse pin through the pendulum, by the traditional-shaped rectangular bottom, or use two impulse pins that act on either side of the pendulum block. This method is described below, as it offers a second option to the slotted bottom previously described. In this instance the impulse pins are steel, and are spaced to accommodate the pendulum impulse block between them.

The photos and drawing show the arrangement for this. Note there is a small part of the rod section left square in shape to accommodate the beat adjuster. This is not required unless you are making the beat adjuster.

The author has also chosen to make the crutch section half-round rather than fully circular. This is to show an alternative slightly stiffer design, as it does not need to bend for setting the beat. Note the steel section left unfiled has a brass cap fitting over it. The section left square is 30mm down from the top hole centre.

Mark-out the steel sheet as shown in the drawings and photos. Begin the construction by drilling the three holes, the top central one of diameter 3mm to match the pallet arbor diameter. Drill the bottom two holes to 1.4mm diameter to allow for tapping for a 10BA thread.

Now, using the piercing saw, cut the shape out taking care to follow the lines carefully. This crutch is flat in its final form and is therefore a filing exercise to shape and finish the edges. To form the half-round rod section of crutch it is not hammered to begin with, but filed. Some initial shaping can be carried out using the vice, but most of the shape is formed by using needle files and placing the crutch on the edge of the workbench to file a quarter-circle on each side. The idea is to keep swapping round until by eye the shape appears as half-round. Take care not to file the section where the beat adjusting screws are going to make contact. To finish the shape the author has used what might at first glance appear to be a very odd technique: holding one end of the crutch and resting the other end against you, it is possible to file it and feel the file cut-

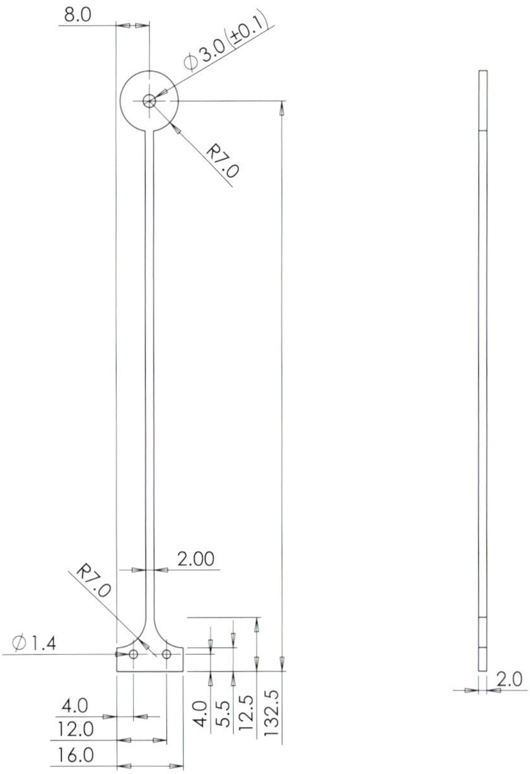

Crutch for adjustable beat arrangement.

Crutch pin.

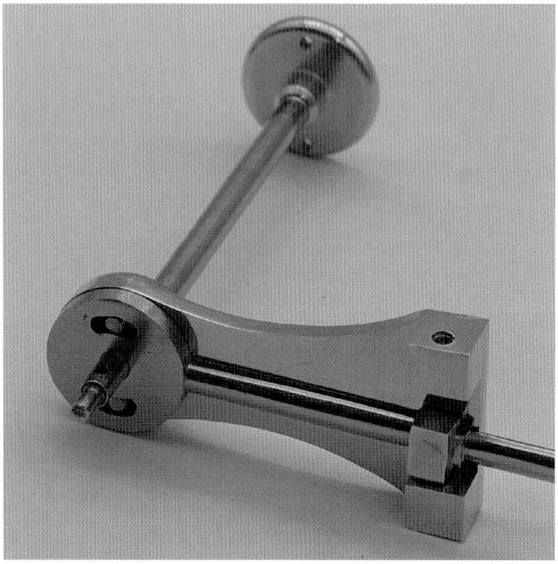

Here the adjustable crutch is partly assembled to show the details of how it functions.

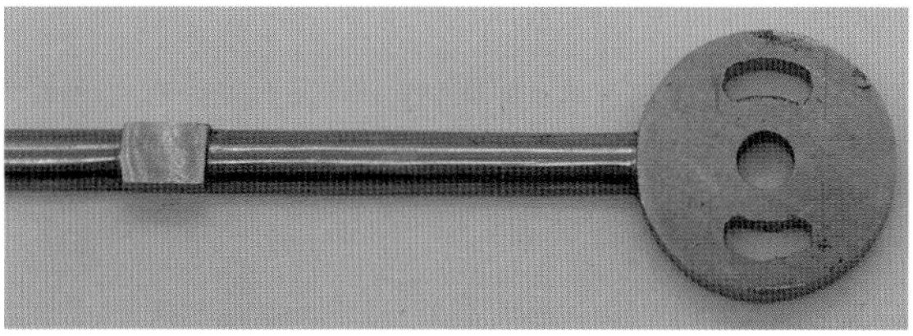

The section of crutch left as a square section for the beat adjustment screws to act on in due course. Note that the curved slots are not required for this project.

ting it to shape. It is not easy to explain, but it does work! This enables you to file right round the half-circle in one go and will help make the finished crutch rod look like a complete half-circle. Finish using the self-adhesive micro finishing gel attached to the file. Work through the grades of finishing gel, gradually getting finer until the finish you require is achieved.

There remains the two impulse pins to turn, using a good-quality steel of diameter 3mm. Follow the drawing to reduce one end to a diameter of 1.7mm in order to use a die to form a 10BA thread on the end. Polish the bar for a length of 50mm before parting in two sections and cutting the threads, as this will avoid catching the threads with a buff stick.

Begin by placing the steel bar in the lathe with 5mm forward of the collet. Turn the diameter down to 1.7mm for a 3mm length. Next, using the 10BA die placed in the tail stock, bring the die to the workpiece. Turn the head stock by hand and cut the thread by turning half a turn forward and back. It takes a little time and care as the thread being cut is small. Use the same oil as used on a piercing saw blade to lubricate the steel as you cut with the die. Finish the ends of the impulse pins and fit them to the bottom of the crutch.

Brass mounting collet

The whole assembly will be attached to a brass collet similar to those used to mount the wheels. In this instance the diameter of the end of the pallet arbor used is 3mm to allow the mounting of the middle section of the assembly. Using the lathe, turn the collet to the dimensions shown and either soft-solder it to the arbor or secure it with Loctite 638 as shown.

The actual beat adjustment

Take a section of brass 6mm thick and long enough to allow the marking-out of the workpiece with enough extra to hold it in the vice. This allows all aspects of the adjuster to be worked on without leaving vice marks. Using the scribe, set square and divider, mark-out the shape shown in the drawing.

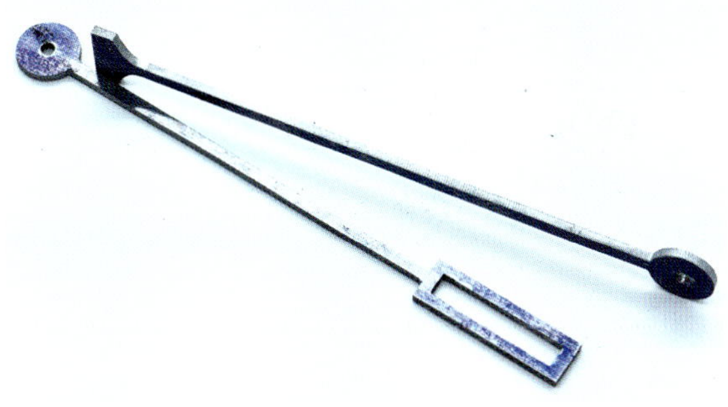

The two variations of the crutch cut with the sides tidied to make them more comfortable to hold whilst working with them.

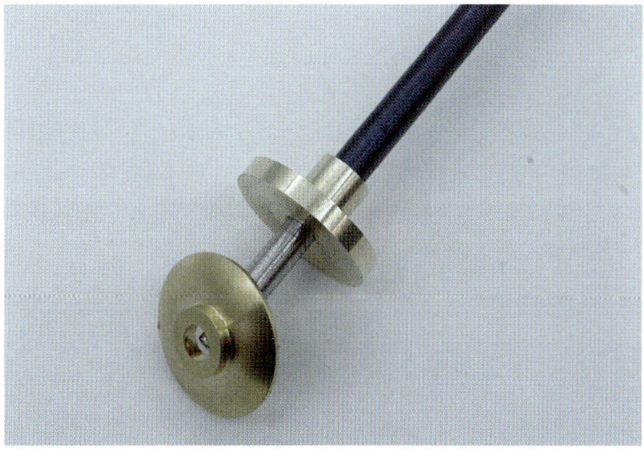

Mounting the pallet collet using Loctite 638.

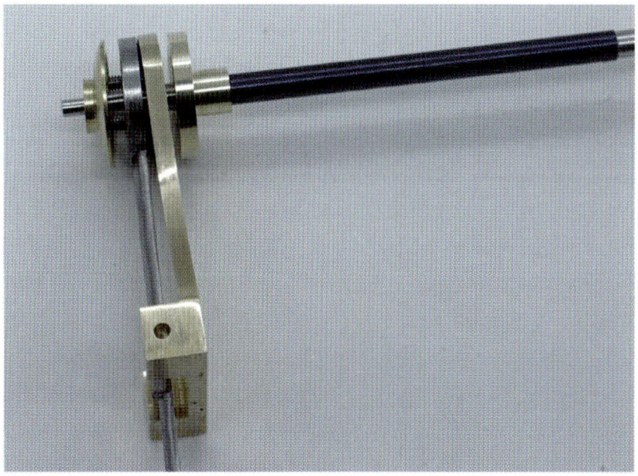

All the parts of the adjustable crutch assembly placed in position but not finally assembled, to illustrate the order of assembly.

This is one of the few parts for this clock where access to a milling machine can be useful. It is quite a precise filing exercise; you need to take care to file with a flat filing action, not rolling or tipping the file. As the author has a milling machine, the example shown has been milled.

It is best to start by drilling the holes needed before filing or milling the workpiece to shape. The location for the holes and their diameters are shown in the drawing.

Drill the central hole to match the diameter of the end of the pallet arbor – that is, 3.0mm in the drawing. During the turning of the pallet arbor you may have finished up with a slightly different diameter. If this is the case, drill this centre hole to match your arbor diameter. The object is to have a closely fitting beat adjuster. Next, drill the holes either side of the centre hole just 1.4mm diameter at this stage. It is now necessary to shape the remainder of the beat adjuster.

Begin by filing across the upper section, taking care not to file beyond the line scribed for the lower section. There needs to be sufficient brass in the bottom section once it is shaped to allow the fitting of 8BA screws for the beat adjustment.

Next, file down between the 'lugs' we require for the 8BA threads. It is a good idea to place masking tape over the section already filed to help protect it from being caught as you file through the lugs.

It is now necessary to apply a finish to the work, and this is perhaps one of the most awkward parts to finish off neatly. It is quite tricky not to end up with all the edges and corners a bit rounded. Use the gel film product stuck on a file here as it is easier to keep this flat and avoid the corners than if using a wooden buff stick. This at least means you can finish one part without touching any other sections and spoiling any previous work. A small oil stone lubricated with white spirit also works very well if you have one.

MOUNTING THE BEAT ADJUSTER TO THE COLLET

Once the workpiece is finished to your satisfaction it is time to consider how to mount it to the collet on the pallet arbor. There are already two holes of 1.4mm diameter drilled to aid this. Place the pallet arbor in the machine vice on the drill table, with the collet uppermost. Put the beat adjuster onto the pallet arbor

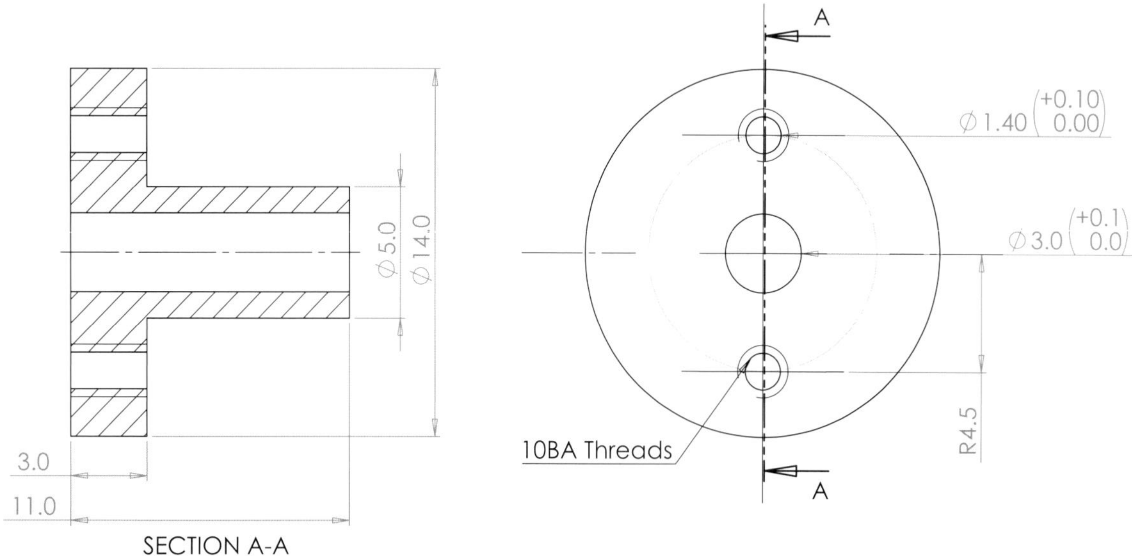

Adjustable-beat collet.

The finished brass base section of the beat adjuster mounted against the collet.

in contact with the collet already mounted. Now using the 1.4mm diameter drill, pass through one of the holes drilled in the beat adjuster, and drill through the collet. Remove both from the vice and tap the collet hole with the 10BA tap to form the thread. Now, drill just the one hole you have used in the beat adjuster out to 1.7mm to give clearance for the 10BA screw. You should now be able to screw the beat adjuster to the collet on the pallet arbor with one 10BA screw. Placing the pallet arbor back in the vice and using the beat adjuster as a guide, drill the remaining hole 1.4mm, remove the beat adjuster, tap the hole 10BA and drill the remaining hole in the adjuster 1.7mm to give clearance. You now have two screw holes that exactly line up to secure the beat adjuster to the collet. Countersink these two holes so that a countersunk 10BA screw sits completely flush with the surface of the beat adjuster. The screw heads must be flush with the surface of the beat adjuster so as not to foul the movement of the crutch.

The one remaining task is to tap the two 'lugs' 8BA to allow screws to be fitted and the beat to be adjusted. Use an 8BA taper tap to begin, take care not to touch the bottom of the section between the two lugs. Use the plug tap to finish the threads.

Securing the components together

A way is needed to hold the crutch into the beat adjuster whilst allowing the adjustment. This is done by drilling a hole of 0.8–1mm diameter through the arbor for a securing pin to pass through. Drill this hole 2mm in from the rear pivot shoulder on the pallet arbor.

A domed friction washer, shaped rather like a minute cap washer – the cap which holds the minute hand in place – is turned in the lathe from a section of brass bar. It can be a plain washer if you wish, but the author prefers to use a shaped one as shown in the photo on

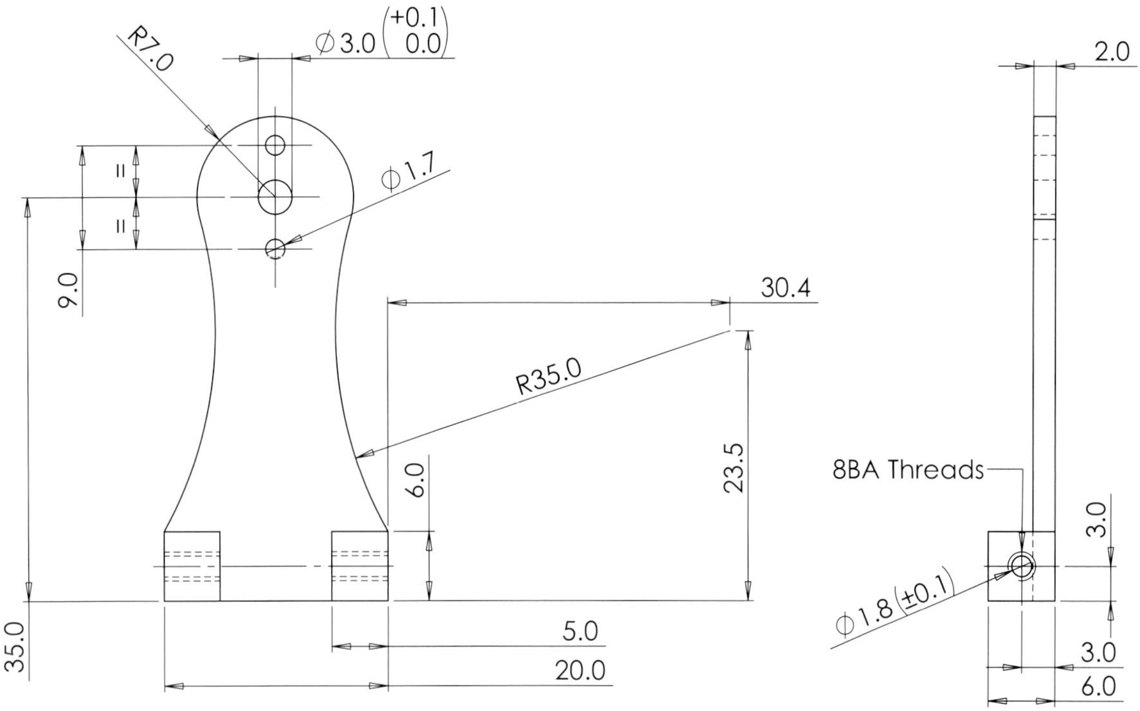

Beat adjuster.

page 143. To turn the domed washer, a stepped collet is a great way to hold it in the lathe. If you do not have stepped collets, make a flat washer rather than a domed one.

Making the domed washer

Begin by facing the end of the brass bar in the lathe, then use a centre drill of less diameter than the finished hole, to mark the centre. Next, drill a hole the same diameter as the pallet arbor end diameter – in this instance, 3mm and approximately 6mm into the brass bar. Make the washer with an outside diameter of 14mm and initially 3mm thick. Shape the end as shown in the photo. The trick with this washer is to finish up with the correct length to secure the crutch firmly but not immovably to the pallet arbor. A bit of careful measuring here will pay dividends. Screw the pallet adjuster to the collet, put the steel crutch into place, and then carefully measure with the vernier gauge the distance between the front of the crutch and the edge of the hole in the pallet arbor. Add approximately 0.3mm to this measurement and then, using the parting tool in the lathe, part-off the washer to the measurement you have. Now reverse the domed washer, placing it back in the lathe with the parted end showing. Turn a shallow angle on this side of the washer, giving a concave surface. This helps it sit firmly against the crutch.

Place the washer onto the crutch and put a steel tapered pin through the hole drilled in the arbor. Hopefully your crutch will be firmly held, but still moveable. If it is too tight, reduce the height of the washer slightly and try again. If it is too loose, make another washer.

Insert the beat adjusting screws and you will see that as you adjust the screws the crutch moves from side to side. This adjusts the beat.

14 Making the Pallets for the Escapement

The recoil escapement pallets.

AS PREVIOUSLY DISCUSSED, the design of the escapement developed through history to give better timekeeping as the quality of materials and science behind the designs improved.

The most straightforward escapement to make, giving satisfactory timekeeping, is the recoil escapement. The escape wheel teeth are straightforward to cut and not too thin. This is the first option.

More of a challenge to construct is the deadbeat escapement, which gives very good timekeeping with a seconds pendulum – that is, a pendulum that beats once every second – such as we have in this design. Several versions of the deadbeat escapement have been developed over time. Four types of the deadbeat escapement are commonly found: the Brocot pin-pallet deadbeat escapement; the pin wheel escapement; and third is a type found in many continental weight-driven wall clocks where the pallet pads are separate steel items set into a brass frame, which allows for adjustment. Fourth, there is also the lever escapement, which is found in the majority of carriage clocks. These often have jewelled pallet pads. (A jewelled pallet pad has less friction than a non-jewelled pad. *See* W.J. Gazeley's book, listed in the Appendix.)

Whilst the traditional version of the escapement gives the best performance, it is quite an undertaking to accurately construct, especially when setting the impulse angle for each pad, and then polishing these and the locking curves. If these are corrupted in any way the pallets must be made again as the performance will be compromised. In addition, the escape wheel must be cut with its teeth offset to produce the necessary profile for the pallets to engage correctly.

The object of this publication is to be able to produce a well-made clock without needing a working background in horology, so for the deadbeat escapement option the author has elected to describe the construction of the Brocot pin pallet escapement. Its performance is very good and it is somewhat easier to lay out and construct. The escape wheel teeth are also radial, so the need to offset the cutter to produce the correct profile is avoided.

The diameter of the escape wheel is 38mm for the deadbeat escapement and 44mm for the recoil version. These sizes have been chosen in order to allow the pallets to be a little larger and therefore more straightforward to mark out and construct.

MARKING AND CUTTING THE PALLET BLANK

Traditionally, having marked-out the pallets on the steel to be used, the area round the marked lines is drilled with a series of holes, cutting down on the amount of sawing required to cut out the general shape; this is known as 'chain drilling'. The author has found that although it does cut the sawing down the amount of finishing required rather outweighs the time saved by drilling. It also looks ugly and has sharp edges, which can make it rather awkward to handle whilst being made.

Using a piercing saw to cut the whole shape out gives a surface that is easier to work with, and much quicker to finish. Having done the sawing exercises and cut the wheel crossings, you will be beginning to master the piercing saw.

Piercing saw tips

The secret to cutting the pallets out is to use a fairly fine saw blade – for example, 0/2 – and keep it well lubricated with a light oil. Almost any thin oil will work to lubricate the saw blade and help it stop clogging. Fit the blade cutting on the down stroke. Take your time: do not push the blade into the steel faster than it will cut. Once the blade loses its edge, change it rather than waiting for it to snap. As can be seen, there is comparatively little finishing to do. The issue with any finishing of the clock movement is trying to keep the edges of your work clean and sharp. The more tidily you work, the less finishing is required, and the less chances of rounding the corners of your work.

The roughly cut recoil pallet blank (1).

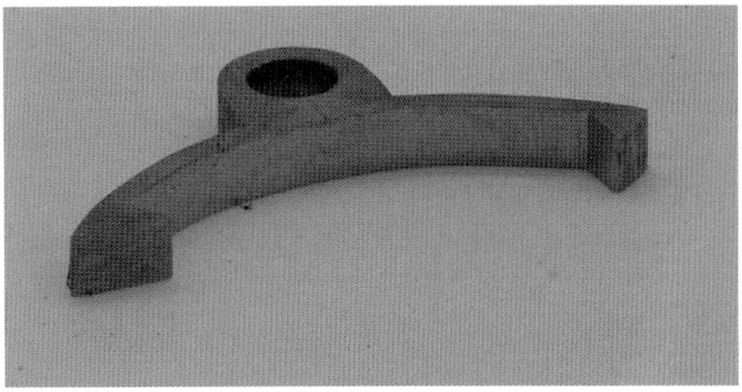

The roughly cut recoil pallet blank (2).

CONSTRUCTION OF A MARKING AND TESTING JIG FOR THE RECOIL PALLETS

The base of the jig marked with both the centre holes for the two different types of escapement.

Marking out the recoil escape pallets is best done by making a working jig. It greatly helps to shape the pallets and get them working, before finally fitting them once the escape wheel is mounted in the movement.

Use an off-cut of brass measuring approximately 50 × 100mm and preferably 5mm thick. (A piece of steel or aluminium could also be used.) Colour it with a dark blue/black colour and mark a centre line down the middle. Next, punch the distance between the pallet to escape wheel centres onto the line marked down the middle of the plate. For the recoil escapement this will be 1.4 times the *radius* of the escape wheel, for this clock 1.4 times 22mm, so 30.8mm. The deadbeat escapement has a different centre distance, rather than 1.4 times the radius of the escape wheel, it is the *diameter* of the escape wheel. When marking-out the jig the centre hole is therefore 38mm for the deadbeat escapement. If making the recoil escapement, turn two stepped steel pegs as shown in the lower section of the photo. The larger diameter matches the centre hole drilled in the escape wheel – 6mm. The smaller diameter – 5mm – is turned to fit the 5mm-diameter holes that are then drilled right through the jig base plate. In the 6mm-diameter end of each peg, use a centre drill to drill the centre point. This gives a very useful reference point when continuing the marking-out of the recoil pallets on the steel blank.

The pegs should be a close fit into the base holes, and the escape wheel be a good fit onto the peg. The holes being drilled right through the base of the jig gives a way of removing the plug if necessary, and avoids an airlock trying to fit or remove the peg if needed.

The jig base plate will work for both designs of escapement, although it is most useful for marking-out the recoil pallets and getting them to work. Bear in mind the different centre distances for each type of escapement: 38mm for the deadbeat escapement and 30.8mm for the recoil escapement.

The recoil pallet material

The pallets need to be made from a good-quality steel, such as BS4659 B01 AISI type 01. This is ground flat stock, a high-quality 'tool steel' that's been annealed – which makes it easier to work and will harden and temper well. This steel is 'bright' – that is to say, not covered in mill scale. (If your steel has mill scale you need to remove this before beginning, using coarse abrasive paper or film.) As with all the other parts that require marking-out, try to keep your marking-out light and neat, as it will all need to be removed – this is a more arduous task with steel than with brass. Once finished, the pallets will need to be hardened and tempered in order to wear well. The dimensions of the steel stock to make the recoil pallets from are 56 × 30mm and 4mm thick.

This steel is readily available; *see* the suppliers list in the Appendix.

MARKING-OUT AND MAKING A SET OF 'RECOIL' PALLETS

A drawing is included with the likely dimensions for the recoil pallets, but it is much better to use the following method to mark them out. It would serve as a useful check to compare your marking-out with the drawing; they should not be very different.

Mark the centre of the piece of steel you are using to make the pallets – the pallet blank as we will now refer to it – with the scribe and steel rule going across the width of the pallet blank rather than down the long dimension. Next, centre-punch and drill a hole through the pallet blank 10mm down the centre line from the top edge, the same diameter as the pegs in the testing jig – 6.0mm. In due course this will be the mounting hole for the pallets.

You will then be able to place the pallet blank onto the peg on the jig and line up the centre lines of the pallet blank and the jig. You could use some double-sided tape to hold the blank in position. Place the escape wheel onto its peg so it sits on top of the pallet blank, with the tip of a tooth sitting on the centre line, as shown in the photo.

Now count four teeth to the left and make a very small mark; label this mark 'C'. Count to the right four teeth and make another very small mark, labelling it 'D'. Try to get these marks as best as you can to show the tip of the tooth. The author uses an 'optimiser' visual aid to help.

Turn the wheel half a tooth to the right. (Take care that the centre lines marked on the base of the jig and the pallet blank still line up correctly.) Once again, count four teeth to the left, mark this point and label it 'E'. Also count four teeth to the right and label this mark 'F'. These points are shown in the photo.

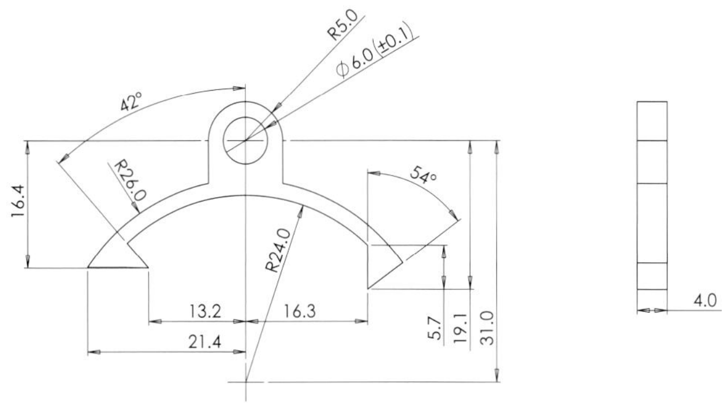

Recoil escapement pallet.

The escape wheel is placed on top of the steel section to be used to make the pallets.

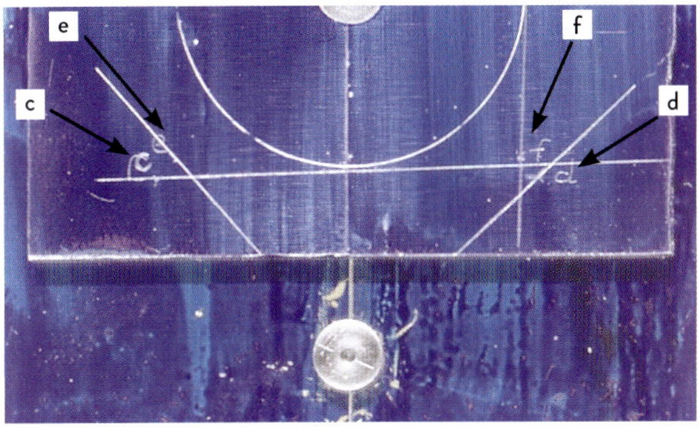

To show points c, d, e, and f marked as described in the text, the escape wheel has been removed from the jig.

Take a set of engineer's dividers and set them to the distance from the centre line to point 'F'. Scribe a circle from the PALLET centre of this radius; let's label the circle 'G'.

A line is then drawn from point 'C' to touch the circle, and will pass mid-way between points 'D' and 'F'.

Scribe a line through 'F' to touch the circumference of circle 'G'.

Scribe two more lines, one from the ESCAPE wheel centre passing through 'E', and another from the ESCAPE wheel centre passing through 'D'.

This gives the shape of both the entry and exit pallet impulse surfaces; the only remaining lines left to scribe are the ones denoting the actual shape of the pallets.

For these, scribe two more lines – one from the ESCAPE wheel centre of radius 24mm and another also from the ESCAPE wheel centre of radius 26mm.

Shown are all the marking-out lines required to form the recoil pallets as described in the text.

Traditionally they are shaped as in the photo. However, you can shape them to your own design, bearing in mind that lightness is good.

Take your time with the piercing saw, concentrating on holding the saw in the vertical position, and following the marking lines closely as you saw.

Having cut the pallets out, use a fine file just to tidy any rough areas to make them easier to handle. Do not file what will be the impulse surfaces at this point.

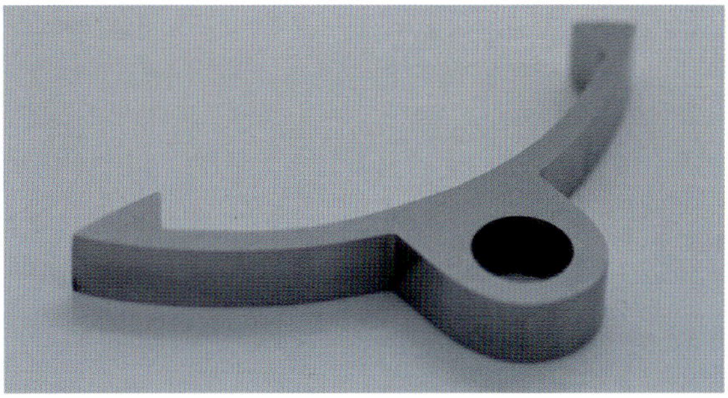

One of the traditional shapes used for recoil escapement pallets.

Using the piercing saw to cut through the pallet blank.

MAKING THE PALLETS FUNCTION

Next comes the process of making the pallets function. This requires some careful filing in the right area. Knowing just where to file is something gained with experience. Care is required not to file too much away. The object of the exercise is for the entry pallet impulse surface to release the tip of the escape wheel tooth, enabling the wheel to rotate 1 degree (the thickness of the tip of a tooth) before the exit pallet impulse surface catches a tooth. The object is then for the exit pallet to release the tip in due course, enabling the wheel to turn 1 degree before the entry pallet catches another tooth, and so on.

Begin by smoothing the impulse surfaces, taking care to *remove as little steel as you can*. Use a fine file for this and try not to let the file clog, as this will score the surface. Draw-file and then use a fine India stone to smooth the surface. An India stone is a type of fine oil stone. Use a fine oil stone if you have one. Work along the impulse surface, not across it. When filing or polishing the impulse surfaces it is important to use the 'draw' technique so that the lines of work follow in the same direction as the action of the escapement. That is to say, the escape wheel tooth will run along the impulse surface, so file along the surface, this will cause less wear on the escape wheel teeth. Place the escape wheel in the jig, and the pallets on top in order to try the action. When cutting the escape wheel three wheels were cut together; use one of the spare wheels you have for this – it saves marking or damaging the wheel already crossed. It is quite likely the escapement will not yet pass; further adjustment is made by reducing either the thickness or the length of the impulse surfaces. Make only very fine adjustments and check often, working the surfaces as smoothly as you can. There will not be enough material to allow for removing deep file marks. It is not possible to say exactly where your individual pallets need adjusting to work. Check often and observe both of the pallet impulse surfaces – the entry and exit – **after each adjustment**. Each adjustment made will alter the *whole* action of the escapement. With careful observation you will see the effect removing just a small amount of steel will have. Check for any burring occurring and remove these as you work.

When the pallets will pass in the jig they need to be mounted on the collet attached to the pallet arbor so they can be tested on the mounted escape wheel. Do not yet rivet the pallets to the collet as they will require hardening, tempering and finishing before being finally secured in place. Mount the pallets and crutch on the pallet arbor and place them in the clock movement with the centre, third and escape wheels. Do not just put

The pallets filed to the extent they will now just pass the escape-wheel teeth.

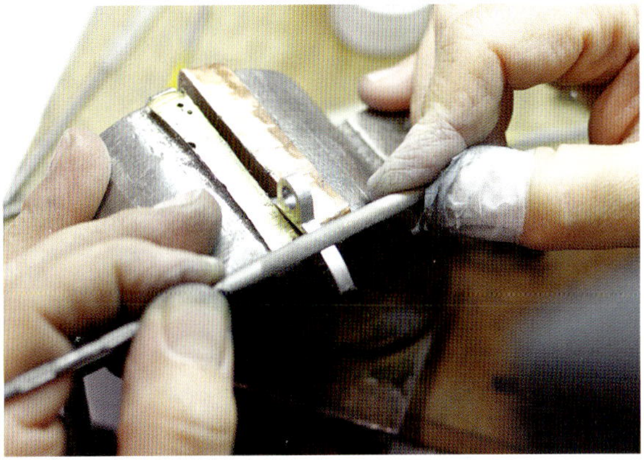

Draw-filing the general shape of the pallets.

the escape wheel in, as pushing this directly will give a false impression and you may damage the teeth of the escape wheel. Now, apply gentle pressure to the centre wheel in a clockwise direction and try moving the pallets back and forth to see if the escapement will pass – that is, work. The action should be as previously described. It will probably catch at this point. If so, proceed in the same way as when using the jig – taking care to remove only the smallest amount of material with the file. It is a question of patience and trying the pallets often.

Once the escapement is working, remove the pallets from the arbor collet for finishing. Draw-file the edges of the whole pallet before using gel sheets stuck to a file to finish and polish the edges and impulse surfaces. Then use the gel sheets secured to the plate glass to finish the sides. The final process is to harden and temper the entry and exit impulse sections of the pallet.

HARDENING THE RECOIL PALLET IMPULSE SURFACES

Having made the pallets, ensured they work, and finished the surfaces, it remains to harden and temper the impulse surfaces. The whole pallet can be hardened and tempered, but only the acting surfaces actually *need* to be. Time removing the heat scale can be saved by only heating the ends of the pallet. Each end is heated individually to red hot and plunged into water. There is no need to heat the pallet to bright cherry red – just to red hot will be sufficient. Wear the appropriate safety gloves, and use long-handled pliers. If in doubt, heat the whole assembly.

Once either end or the whole pallet has been hardened, it then requires tempering. Clean the fire skin from one side of the pallets using a reasonably fine gel sheet on the plate glass. Once the bright steel is visible, reheat the whole pallet gently until it turns the colour of straw. Quench in water at this point to fix the temper. Were you to carry on the steel would turn dark straw, purple and blue, at which point it may have gone a bit too far and begun to soften slightly. If so, the process will need to be repeated as soft impulse surfaces will wear quickly.

MARKING-OUT AND MAKING A SET OF BROCOT-TYPE DEADBEAT PALLETS

The Brocot type of deadbeat pallet is constructed from a brass or steel frame and two circular steel pads. The drawing shows the shape and dimensions for a frame of a fairly traditional style.

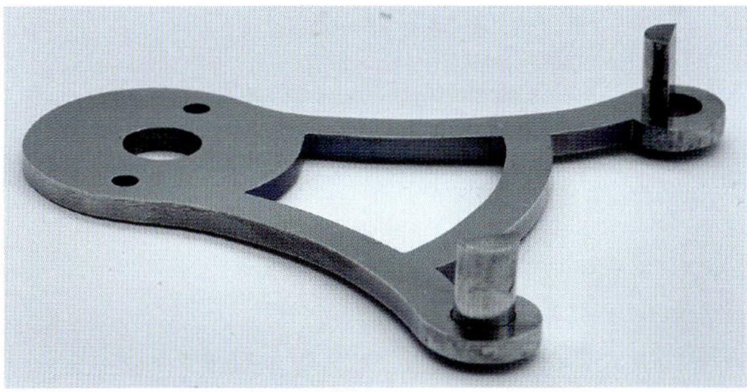

The steel frame shown cut out and finished well enough to have a trial fit.

The finished pallet assembly.

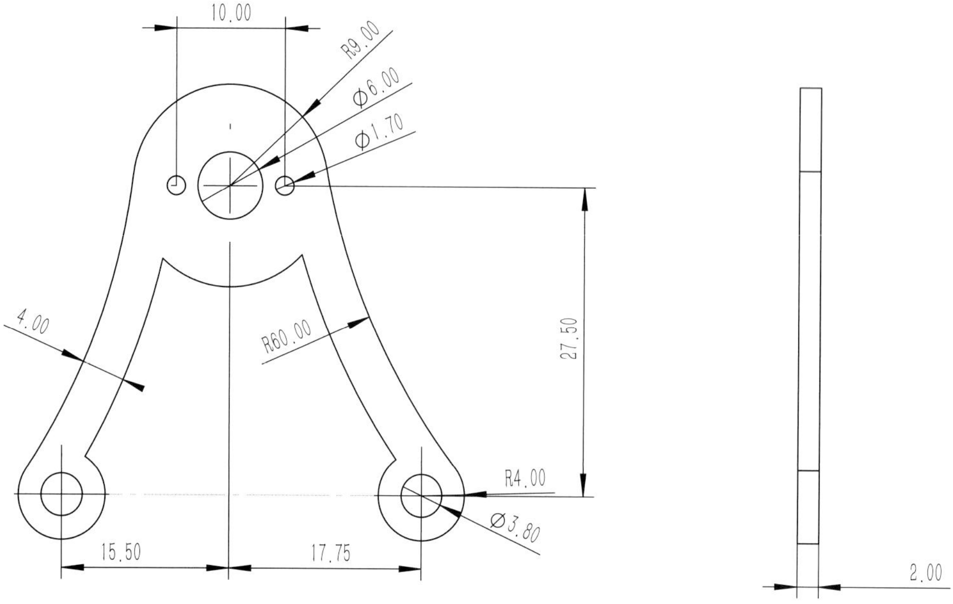

Deadbeat pallet.

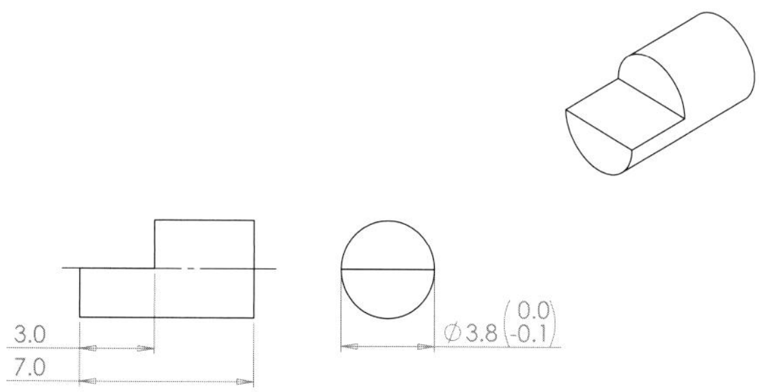

Deadbeat impulse pad.

The impulse pads are made from blue steel – known as *pivot steel* in the horological industry – and is supplied already hardened and tempered. It can be purchased from the horological suppliers listed in the Appendix. If you are unable to source pivot steel, use EN24 as this will harden well.

The steel frame is considerably easier to cut and shape than the recoil pallet frame, as it is thinner and not relevant to the acting geometry of the escapement. The holes for the centre of the pallet frame and the two holes for the impulse pads to be fitted can be drilled before cutting out the frame. Since it is critical that the correct size holes are drilled in exactly the right place, if you miss-drill them begin again rather than trying to correct them.

The pallet pads

The diameter of the impulse or pads should be chosen so that as a complete diameter they will fit between the teeth of the escape wheel to a point just lower than the centre of the pad. If the diameter is larger than this, the escapement is unlikely to function – and if it did the locking would not be safe. If the diameter is smaller it will function but the excessive drop would mean it would be a poor escapement. The diameter of the pad used by the author and shown in the drawing is 3.80mm. Pivot steel is available in this diameter.

Having drawn out the pallet frame, centre-punch the three positions – one for the centre of the pallet arbor and two for the impulse pads. Take good care here as these holes must be correctly located. Drill the arbor hole to be a close fit on the pallet arbor.

Ideally, the impulse pads will be a tight push-fit into the pallet frame. It is worthwhile drilling a test hole in a piece of scrap brass to check that the diameter of the steel to be used to make the pads is a tight fit. Check this fit before going further, as it might be that the pins will be loose in these holes. If so, drill another test hole 0.1mm smaller and try this fit. The idea is to be able to tap the pins finally into position when all is adjusted without the need for Loctite 638, shellac or solder to secure them.

Having selected a suitable drill for the impulse pads, drill these two holes through the frame. Take special care

Steel known as 'blue' or 'pivot' steel in the horological industry.

with these holes – the correct action of the escapement relies on these holes being upright and in the correct location. As discussed earlier, the actual shape of the frame can be your own design. Bear in mind that the oscillating parts of the escapement work best when they are as lightweight as practical. Part of the design of the pallet frame must also allow it to be secured to the arbor. As with the wheels, it can be either riveted or screwed. Shown is the frame designed to be screwed against the collet, which has been soldered to the pallet arbor. For pallets the author's preference is for the pallets to be removable. Two screws are usually sufficient to secure pallets to the arbor. A brass collet is required to screw the pallets against. Turn it to the dimensions shown.

It is quite in order to go ahead and finish the frame at this point, as once the pads are adjusted and tapped into place it is harder to finish the frame, and unwise to remove the pads again. The frame does not need to be hardened and tempered.

Making the steel impulse pads

Blue steel – shown above in the form it is usually purchased, and known as pivot steel in the horological industry – is used for the impulse pins. It is a good-quality steel that polishes well, and also hardens and tempers well. It is also purchased 'true', so can be used straightaway without the need to turn it true. The finish on the steel is also very good and so little final finishing is required, provided you work neatly.

The two important aspects of making the impulse pads is that they must be the correct diameter, and that the end used for the action of the escapement is filed to exactly half the thickness of the diameter. Too thick and the escapement will not pass; too thin and the escapement will be compromised. They must also be of constant thickness, not tapering, as this will affect the action of the escapement when the end shakes alter position. It is necessary to make these pads well. If your first attempt is not very good, have another go, as it is just the pin you need to re-make.

Cut a piece of pivot steel off about four times the length needed to make both the pads, about 30mm. To cut it, either turn it in the lathe using a hand graver or a thin-tipped tool. Alternatively use a file, as a saw blade will generally not cut it.

Place this short section of steel in the lathe with about 10mm clear of the collet and turn a groove until there is about 0.5mm left in diameter. It is not an exact measurement. (The idea is that when the work is further on in operation it is possible to snap the pad away from the steel at this point.)

Tidy the end of the bar, facing off and polishing it. Reverse the length of steel in the lathe and repeat with the other end. Next, polish each end section diameter. This is done now, as if done when the pads have been formed there is the risk of rounding off the edge of the impulse pin. Start with 1,200 grade or gel sheet equivalent, as the steel should have quite a good finish as it comes from the supplier. Move through the various grades to 2,000 grade paper and then to Autosol polish used on a piece of hardwood.

The section of pivot steel used to form the pallet pads. Shown here with both ends marked ready to reduce the diameter to approx 0.5mm for separation in due course.

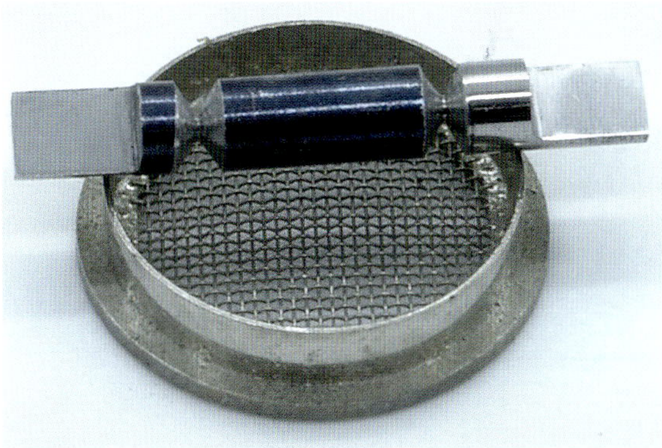

The groove cut into the steel ready to be snapped off once the shaping is complete. It also shows one pad shaped – left-hand end – and one pad finished – right-hand end.

You now have a piece of steel 30mm long, with a groove turned in 10mm from each end, polished.

Filing the impulse pin to shape

The steel is held in the pin vice in such a way as to be able to file it, but not damage the end of the pin vice jaws. Place a couple of brass washers on the steel so it comes between the safe edge of your file and the pin vice. Leave only the length required for the acting impulse pin uncovered by the washers.

Rest the pad to be filed onto the wooden block used for filing pins and file straight across the steel. To begin with, hold the pin vice firmly and establish a flat section across the steel. Once this is done, hold it more lightly and allow the file to find the flat.

With practice you will find the steel will file flat. It is another of those peculiar horological techniques you learn as an apprentice that actually works quite well when you master it. Try an experimental piece before filing the actual pallet impulse surfaces. You will find it will sit in the groove on the block and actually come up quite flat. Use a fine file, making regular checks as to the thickness of the rod. It is necessary to finish with exactly half the diameter of the steel rod, so use a micrometer if you have one; if not, read the vernier gauge as carefully as you can.

This flat back edge is for clearance only and is not used in the action of the escapement. Whilst it does not require polishing, the edge must not have a burr or any roughness, as this will interfere with the action of the escapement and wear the tips of the escape wheel teeth. Finishing it with a fine India oilstone works well, and to add a final finish the author used a fine flat file with a fine self-adhesive finishing gel added to the surface, rather like a steel buff stick.

Place the section to be filed onto the wooden block. Note the file has been lubricated to help avoid the teeth clogging. Were they to do so, it would cause a groove in the flat surface. Whilst this surface is for clearance only, a groove going right across would give a damaged dropping-off edge.

Hardening and tempering the escapement impulse pads

The pads can then be hardened by heating them to red hot and plunging straight down into either water or oil. Do not use a direct fierce heat as it heats the steel too quickly. For small parts the hardness will be more even if the steel is heated more slowly. Place them on a brass tray and heat until they are red in colour. There is no need to heat them until they are bright cherry red – just obviously red will

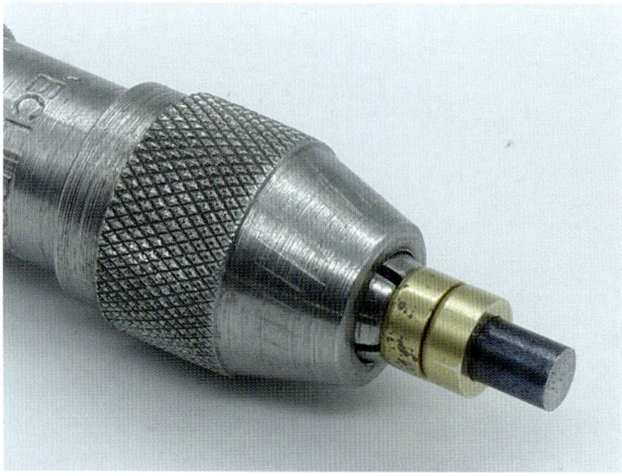

To avoid damage to the pin-vice jaws, and ensure the correct length of steel is filed to form the impulse section, two brass washers have been placed over the pivot steel.

Finishing off the clearance side ensures a smooth dropping-off edge and looks a better job.

suffice. There is quite a lot of discussion as to whether oil or water should be used to quench steel when hardening. Oil hardening is said to work better, as the cooling is a little slower. This can help stop the distortion of the work. The author used water in this instance, which gave satisfactory results. Another technique to harden the steel and also help stop a fire-skin forming is to place the pads in a steel container and cover them with borax. The whole thing is then heated to red heat and quenched in water. Using this method also helps slow down the speed at which the steel becomes red hot.

At this point you have two hard steel impulse pins, one on each end of a piece of pivot steel. Use a buff stick to brighten the surface ready for tempering the steel, taking care not to touch the acting ends of the pins already shaped.

Next, the pads must be tempered, as they are very brittle at this stage. Take some of the brass filings saved from the various filing operations carried out so far and put them in a copper or brass tray ready to apply heat to the bottom of the tray. (Do not use filings that may be contaminated with white spirit.)

Now, place the two pallet pads on the brass filings and apply heat to the bottom of the copper tray. As the heat travels through the brass filings the colour of the steel will begin to change. To begin with it will appear slightly more polished, then it will begin to turn a light straw colour. At this point remove them from the tray and quench them in water to fix the temper. Were you to carry on the steel would turn dark straw, purple, blue and finally grey. Once the steel has gone past the straw colours it would probably be tempered too far and would need to be rehardened. Whilst the pins are still attached to the central section of steel, re-polish them in the lathe. Care must be taken not to round the corners where the curved impulse surface meets the filed clearance, as the action of the tip of the tooth leaving the pallet pad needs to be clean. The impulse surface does need to be well polished. Also tidy the filed flat clearance side.

Finally, the two pins can be snapped off the central section of pivot steel. Place them in the lathe with just the snapped end showing and turn or file the nipple from the end. Polish this end to be slightly domed. Take care not to snap the impulse pin off where the flat surface joins the body of the pin.

The final placing of the impulse pads into the frames requires some care, as the flat side of the pad must not come into play in any way – it is there for clearance only.

As each tooth comes into action its first point of contact should be on the highest point of the impulse pad. Push

The pallets assembled, fitted to the clock movement to illustrate the correct angle for setting the impulse pads.

the pads into each mounting hole far enough to hold in place, but not so hard that they cannot be carefully rotated.

Set them into the frame as near to this angle as you can. Do not push them finally into position at this point, as they may require rotating slightly to adjust the escapement. Mount the pallets and crutch on the arbor and place in the clock movement with the centre, third and escape wheels.

Apply a gentle driving force to the train by pushing the centre wheel in a clockwise direction and observing the action of the escapement. All being well the escapement drop will be even. 'Drop' is the term used to describe how the escape wheel teeth move through the pallets. There are two pads – the entry and exit. For the escapement to be correctly adjusted each pad, or side of the escapement, needs to have the same amount of movement. Turning the pad will alter the point at which the tooth falls off the pad edge, enabling the wheel to turn and the opposite pad receive the next tooth.

Once the drop is even, each pad can be pressed finally home to be secure in the pallet frame.

In the example shown in the photo above, the pallets and escape wheel have been mounted on the outside of the front plate, in order to illustrate the shape and action of the escapement more clearly. The mounting between the plates is exactly the same. The same collet is used, set on the arbor in the same location as the recoil pallets. The pallets are screwed to the collet in the same way as the traditional crutch design discussed earlier.

15 The Pendulum

MUCH HAS BEEN WRITTEN about the pendulum and the many variable points that alter how regularly it oscillates and therefore how well the clock keeps time. Here we will make a pendulum that is relatively easy to source the materials for and will be a good compromise. The suspension spring is the usual we would expect to see, a flat length of spring steel, 0.3mm thick, and readily available from horological suppliers listed in the Appendix of this book.

Pendulum rods have been made in many ways and from many materials, from a simple steel rod, to complex grid-iron pendulums as designed by John Harrison. Carbon-fibre rod is not a bad material, and is cheap and readily available. Quartz rod is a little more expensive and fragile, but makes a good pendulum rod. In fact, a traditional longcase clock pendulum works well, and is quite straightforward to construct. The traditional brass-fronted lead bob is available from an horological supplier if you wish to have the traditional look. For further reading on the subject, Robert J. Matthys book *Accurate Clock Pendulums* has much research into the materials used in pendulums.

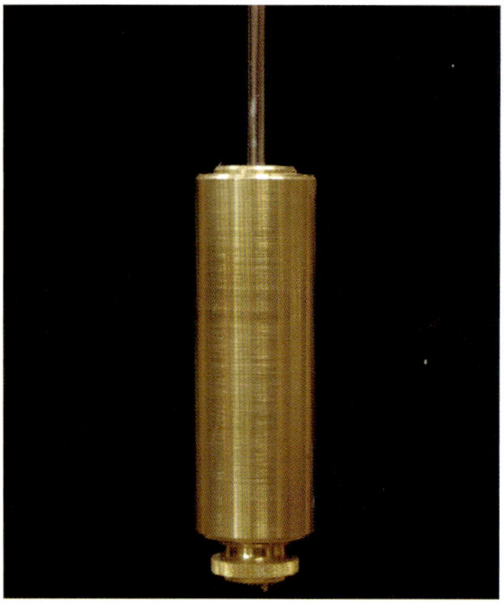

A cylindrical pendulum bob. Shown is the one used in this publication.

MAKING A CYLINDRICAL PENDULUM BOB

A simple type of bob to construct is a cylindrical bob. Whilst it requires more energy from the clock train to drive, it avoids the issues of not having a pendulum bob exactly in line with the direction of travel. Making the whole pendulum adds to the satisfaction of having made the whole clock.

The author has used a variety of materials to make a cylindrical pendulum bob – steel, brass and aluminium silica bronze. For ease of machining and availability of materials, a brass bob is shown for this project. It is a basic turning exercise using the three-jaw chuck. Take care not to mark the bob with the jaws of the chuck. Begin with a piece of brass that is long enough to turn the complete bob, at least 120mm long. *See* the drawing.

With the brass held as far as possible in the chuck, carefully begin the turning operation by facing the end of the brass bar. Next a centre drill is used to drill a hole into the end of the brass bar in order to locate a revolving centre mounted in the tailstock. This will steady the work and give a safer and better quality of finish when turning the bob.

Move the brass forward in the chuck so there is sufficient material clear to make the pendulum bob – 96mm. Using the revolving centre in the tailstock to support the work, the outside diameter of the bob as far as the chuck jaws can be turned, as shown in the photo.

The dimensions for the bob are shown in the drawing on the following page. Part the bob from the bar the correct length and lay it aside for a moment. Make a set of 'soft jaws' for the chuck: they do not need to be a smart finished job, just three strips of brass the length of your chuck jaws, cut using tin snips from a piece of brass sheet around 0.5mm thick. Alternatively, you could cut a strip to wrap round the brass bob and then clamp it into the chuck.

Using the soft jaws, re-mount the bob as far into the chuck as you can with the parted end showing and face this off. We now have the outside diameter turned and the two ends faced.

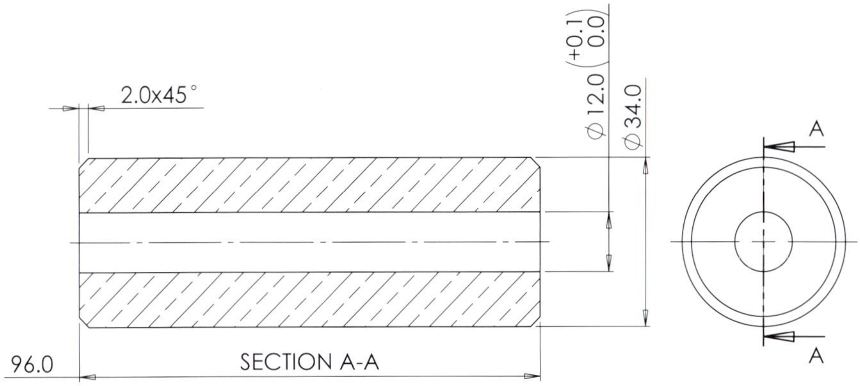

Pendulum bob.

The author's well-used revolving centre.

Drilling the hole through the bob

There remains the task of drilling a hole through the centre of the bob to accommodate the rating section of the pendulum. For this pendulum the rod itself is the rating (discussed in the following text), so the hole drilled through the bob would be 5mm in diameter. We meet a practical problem here in that a 5mm drill will not be long enough to go all the way through the bob. There are several options to consider to overcome this: you could make a shorter bob, which would either be lighter or require a larger diameter to keep the same mass – which would not be practical in this instance, as it would foul on the weight.

The three-jaw chuck can be seen left, with the brass strips acting as a soft jaw visible between the jaws of the chuck and the brass bar stock.

Alternatively, the bob could be reversed in the chuck and drilled from both sides. However, using the chuck rather than a collet to hold the bob, the holes would be very unlikely to be in line, and the drill will grab as it meets the opposite hole, so best to avoid this.

Another alternative would be to use a metric long series drill, which would be available for this operation. The diameter of the pendulum rod is 5mm, so provided there is clearance for the bob to slide on the rod, the hole through the bob need only be 0.2–0.5mm bigger in diameter than the rod. Care is needed drilling a hole of this length and diameter, as using either a blunt drill or too much pressure will cause the drill to go off-centre as the hole is drilled deeper. The pendulum bob needs to sit true to the pendulum rod. Care also needs to be taken to keep the drill flutes clear.

The drawing and photos show a further option, which is to drill the hole through the bob to a larger diameter and then make two end caps to allow the pendulum rod to be a close fit on the caps, and the caps are a press-fit into the bob. The centre hole for the bob shown is drilled to 12mm in diameter, as the 12mm drill will be long enough. With care this will go right through and remain reasonably true.

Two brass end caps are made to fill the gap between the rating and bob. These are shown fitted to the bob in the photo. As these are separate fittings, they can be quite decorative and make an attractive top and bottom to the bob. Try using the round-nosed tool in the lathe to give an attractive radius to the top of the fittings. The round-nosed tool is shown in the photo. Take care with the dimensions for the end caps, as they need to be a tight fit into the bob and allow the assembly to slide freely on the pendulum rod for

The pendulum bob in the lathe, protected from any marking by the soft jaws, with the 12mm-diameter drill being used for the final drill through.

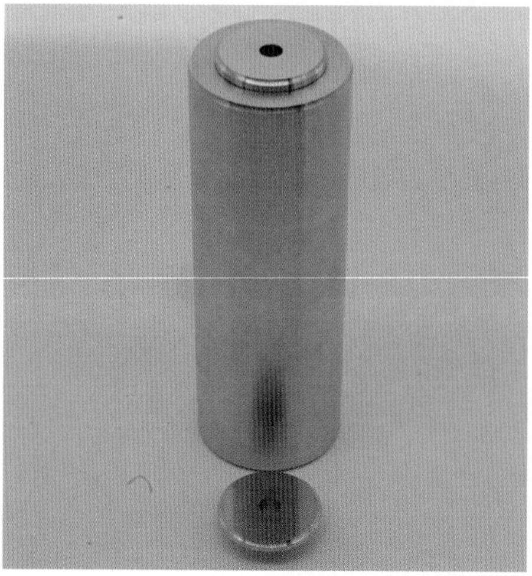

The finished bob with the top and bottom caps.

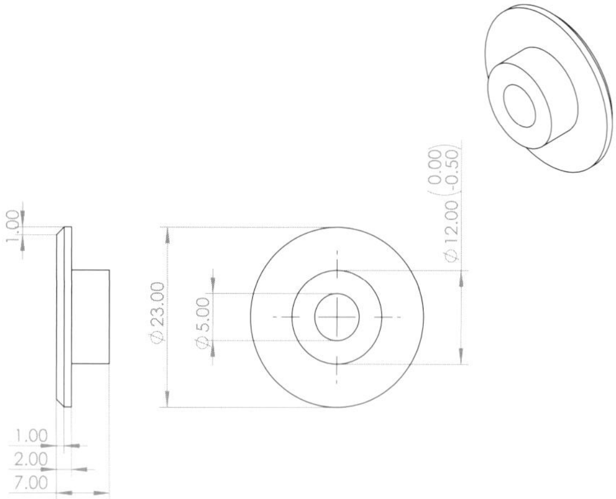

Pendulum bob brass end cap.

regulation, but not move when in operation. As stated previously, note the rating is the bottom section of the rod, rather than a separate component.

When drilling through the bob, begin with the centre drill and work through the drills beginning with 4mm, in 1mm steps using a slow speed. Do not drill beyond the depth of the drill flutes and ensure they do not block with chipping as you drill. Remove the drill often to check the flutes are clear. As the diameter of the drill, its length, and therefore the depth of the hole increases, the depth will go beyond the length of the feed screw in the tail stock. This will require the tailstock to be moved forward. When re-inserting the drill take great care not to 'bump' into the bottom of the hole with the drill. Each time you go up a drill size, the hole will be drilled a small amount deeper. This is not a job to rush. As you reach the end of the hole it will be quite a large drill passing though the end of the brass bob. Take care to drill slowly to ensure a clean cut.

Once the hole is drilled, use a hand-held countersink and, turning the head stock by hand, remove any burr from each end hole. This needs to be done in order for the end cap to sit flush with either end of the bob.

The rating

The rating is the point at which the regulation for the timekeeping of the clock is adjusted. It enables the bob to be raised and lowered by a small amount to regulate the timekeeping. Raising the bob will quicken the timekeeping, and lowering the bob will slow the clock down.

For this pendulum the rating thread is cut directly onto the pendulum rod, which is 5mm in diameter, so it can be directly threaded using an M5 die. There are two choices for the pitch of M5 thread. If you have one, use a metric fine thread, as it will give a finer adjustment. The die required is therefore M5 with 0.5 pitch. You will also need the taps to match for threading the inside of the rating nut.

Since the thread will be cut by hand, pass the pendulum rod through the headstock of the lathe with 40mm or so through and clear of the collet. Do *not* turn the lathe on for this operation. Place the M5 metric fine die in the tailstock die holder, add some oil to lubricate and cut the thread as on previous occasions.

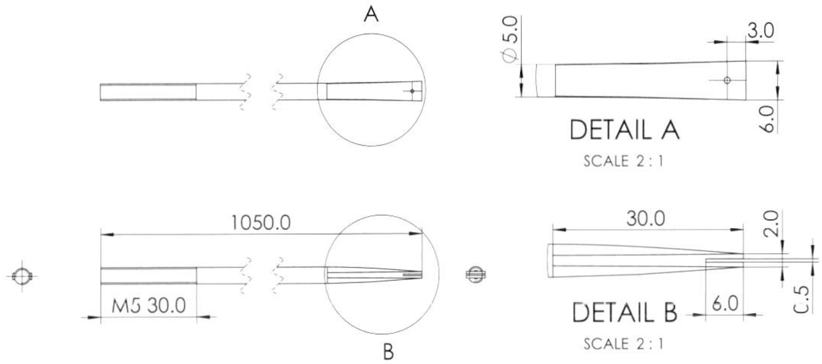

Pendulum rod, solid.

The rating nut

For this operation a knurling tool is useful in order to make it easier to grip the cylindrical nut when regulating the clock. Alternatively – and as they are traditionally made – a square nut can be made from 3mm brass sheet. If making a square nut – as the bob is cylindrical – a spacer will be needed to hold the bob away from the nut to make it easier to adjust.

To make the square nut mark a square with sides 15mm long, and scribe a line diagonally across from corner to corner; this gives the centre of the square. Centre-punch and drill this hole to diameter 4.5mm for fine (4.2 if using 0.8 pitch die) and use the M5 fine taps to thread the hole. Use the piercing saw to cut the square out and finish the edges in the same way other edges have been treated.

A brass spacer is simply a brass tube that holds the bottom of the bob away from the rating nut to aid adjustment. Turn a piece of brass bar to diameter 12mm for a length of 10mm and, using the usual techniques, drill a hole 5.2mm diameter through the centre. Polish the edge, part from the lathe and tidy the ends by facing them in the lathe and finishing with a buff stick.

The round knurled rating nut

The round rating nut is made by turning the profile shown in the drawing into a piece of brass bar of diameter 16mm. Change the cutting tool for the knurling tool and clamp the brass bar between the wheels of the tool.

Start the lathe at a slow speed and, along with a strange sound, the wheels of the tool will turn. In doing so the knurling tool wheels cut a series of interlocking lines into the surface of the bar. Stop the lathe without removing the tool. Once the rotation has stopped, loosen the knurling tool and ensure the wheels are clear of the work before removing it.

Next, drill a hole of a diameter of 4.5mm in the knurled bar and tap M5 fine to match the rating thread. (Drill this hole 4.2mm diameter if using 0.8 pitch.) Whilst the nut is still attached to the bar, finish the end using buff sticks. Part the nut off and tidy the remaining end.

The photo at the beginning of the chapter shows the finished bob, the two end caps, and the rating nut, fitted to the rod.

The suspension spring

To finish the pendulum requires a suspension spring and mounting point. The author suggests a commercial suspension spring is used as the starting-point for the spring. These are readily available. It will need to be shortened and you may wish to replace or re-shape the brass head. Generally suspension spring steel is quite tricky to saw and drill holes through and so it is best to avoid both these ways of cutting. To 'cut' the spring to length, place it in the vice and bend it to 90 degrees – it will snap.

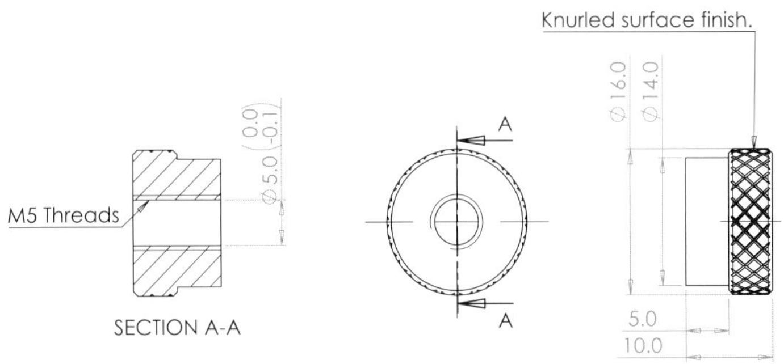

Regulating nut.

The knurling tool in operation.

The effect of the knurling tool.

An oil stone can then be used to tidy the end. When it is necessary to make the hole through it for the two securing pins, use a circular flat-ended punch and matching stake. A stake is a steel base with a hole of a slightly larger diameter than the punch head. The steel spring is placed on the stake with the point where the hole is to be punched through over the hole in the stake. The punch is lined up and then tapped smartly with a hammer. This causes a neat round hole to be punched through the steel.

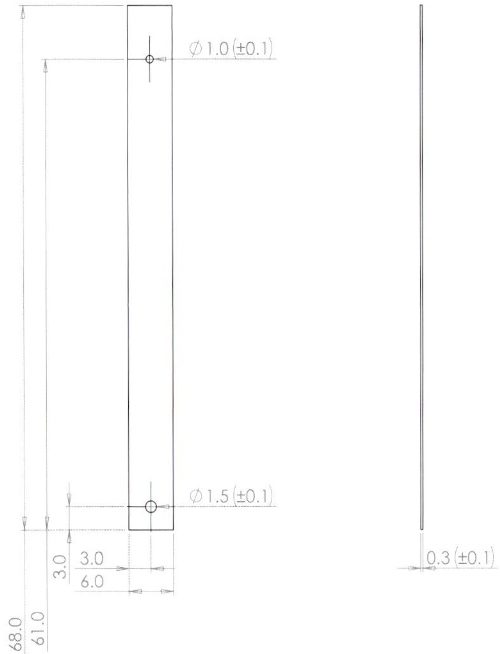

Suspension spring.

If you are using a commercially purchased suspension spring it might well have a brass top riveted in place similar to the one shown in the drawing.

If so, use a hand file to tidy up its general shape and appearance, then measure the correct distance down and break it to length. If making the top of the suspension spring, begin with a piece of brass 2mm thick. Mark out the shape onto the brass as shown on the drawing, in such a way as it allows you to cut the rounded bottom section and the sides of the shape, whilst leaving the workpiece attached to the stock metal. This is so there is a way to hold the piece to cut the slot. Centre-punch and drill the hole for securing the brass top to the spring before you cut the slot. Actually cutting this slot is quite a tricky job, so cutting it before you spend time shaping and finishing it means you will not have spent time only to have to begin again due to a poorly cut slot. Once again patience pays here. There is then the issue of shaping and finishing the rest of the block without closing the slot. Place a section of suspension spring in the slot before proceeding to ensure this does not happen.

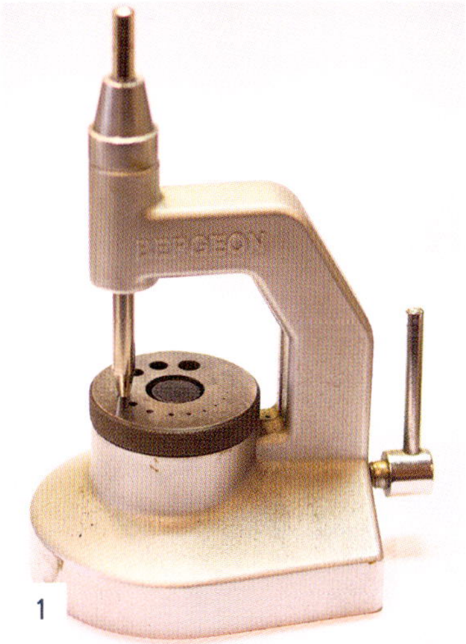

1

A commercially available stake.

2

The punch in line with the correct hole in the base of the stake.

4

3

The suspension-spring steel lined up ready to punch the hole.

The punch hammered through the steel spring.

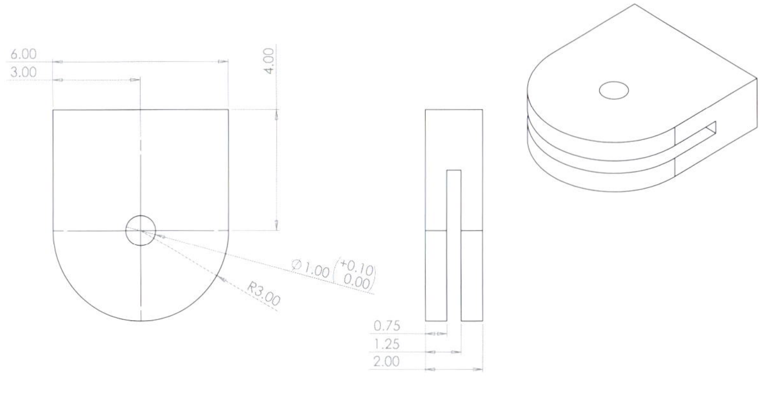

Top of suspension spring.

FORMING THE PENDULUM SPRING MOUNT FROM THE HEAD END OF THE ROD

There are two methods usually found for fitting the suspension spring to the pendulum, either by having a separate brass fitting or, as here, shaping the top of the rod to accept the spring. Here a little blacksmith skill is required. Heat the top 40mm or so of the rod to a dull red heat and, using an engineer's hammer and the anvil, flatten the end of the rod to look as the photograph. This is the only time in the whole project where a larger hammer is used with any force, and after cutting all the thin slots it might come as a welcome relief! Use a file to form the basic shape and then mark, centre-punch and drill the hole for the pin at 1.5mm diameter. There is then one last slot to cut in the end of the flattened steel section.

Once the top of the rod has been shaped, the whole rod then needs to be finished and tidied. Use the buff sticks and gel sheets for this. A good final finish can be achieved by using 0000 grade wire wool and 'Autosol' cream polish.

There is one further small part to make, which can be seen in the photo of the rear of the clock movement with the pendulum fitted.

The diameter of the pendulum rod is 5mm, and the width of the rectangular slot in the traditional crutch is 6mm. This allows for a brass tube to be slipped over the tube so the impulse is given to the pendulum via the brass tube. In use this will make the clock a little quieter and will help the point of contact in the crutch not to wear so quickly.

The dimensions for the brass tube are: outside diameter 6mm; inside diameter 5mm; and length of tube 25mm. Polish the outside of the tube to finish. Slide it up the rod to the position shown in the photo and secure it in place using Loctite 638 or similar.

The suspension spring can now be fitted by placing it in the slot of the rod and passing a steel pin through the hole in the rod to secure the spring. Quite some time ago in the project there was the task of filing steel pins to learn the use of the hand file and pin vice. One of these pins should fit well here.

The bob with its end caps can be fitted onto the bottom section of the rod, the rating, and finally the rating nut fitted to complete the pendulum.

The top of the pendulum following heating and hammering to flatten and widen the end.

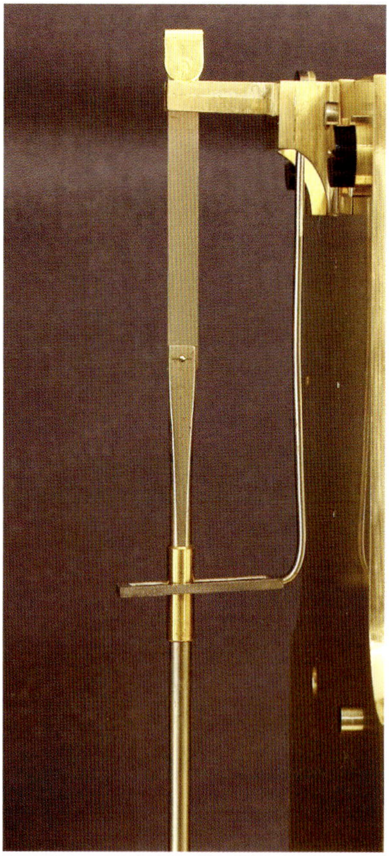

Position of the brass sleeve shown fitted to the pendulum rod, passing through the crutch.

Finishing the Movement

16

WHEN FINISHING THE MOVEMENT parts, decide on the level of finish required. Whilst a very brightly polished movement may be desired, an honest appraisal of the general standard of your work might save a disappointment. Movements look at their best with a definite uniform finish. If the flat parts of the plates are rather scored or the corners are not very sharp, it might be better to give a 'satin-' or grain-type finish.

The more the various parts are polished the greater the risk of rounding the corners of your work. There are a number of traditional ways to polish movement parts. Described is a practical way of achieving a good finish using modern, readily available products. With care you should finish up with the corners sharp and the flat surfaces flat.

The movement constructed with the facing plate surfaces finished to a fine satin or grained finish, and the edge of the plates polished.

Another of the author's movements with the plate surface polished.

USING WET AND DRY PAPERS AND FINISHING GEL SHEETS

Polishing papers and gel sheets for metals are readily available in grades from 120 to 2,000. There are coarser and finer grades occasionally seen, especially coarser for cabinet work.

The gel type of finishing sheets go considerably finer, in fact to a grit grade equivalent to 500,000. For this project a grit equivalent of 14,000 is used. When the self-adhesive finishing gels are stuck onto plate glass, it is possible to retain the sharp corners of your work as you finish it. These are sometimes referred to as 'films' rather than gel sheets. The author recommends using the gel sheets, as the finish obtainable is very good.

Use papers and gels/films that are usable wet as well as dry, as using them wetted (with plenty of water) will extend the life of the paper, but more importantly give a better – cleaner – and quicker finish. Provided your work has been of sufficient quality you may choose to go on and polish the plates. This is done by hand, and not using a polishing motor. Using such a machine would have a tendency to 'pull' the surface of the brass; the direction of the rotary mop is visible when the light catches the surface of the plate. In addition to this, the mop tends to pull the holes and oil sinks into an oval shape.

For all smaller parts, a good finish and sharp edge can be achieved using the self-adhesive gel papers stuck to a sheet of plate glass – use these wet.

Throughout this publication, buff sticks have been used to tidy and finish work. These are best made from an old smooth file or flat brass stock approx 4–5mm thick and 25–30mm wide, with self-adhesive film attached to them. This will help in finishing your work and keeping the surfaces flat.

One of the finer grades of self-adhesive gel sheet secured to a piece of 10mm-thick plate glass. This keeps it flat, helping to keep sharp edges to your work.

FINISHING THE PLATES

To begin the finishing process for the plates you firstly have to decide which type of finish the edges and facing surfaces the plates will have. If the plan is to polish the edges, this needs to be done before finishing the faces. Generally in most 'run of the mill' antique long-case clocks the movements were not polished; instead, the edges and faces were given a uniform bright but matt finish. This is an option for this project, of course; a similar finish can be achieved by using a soap-filled wire wool pad such as a 'Brillo' pad, available in the UK. (Other soap-filled pads are available.) For this, the edges and facings are finished to 800 grade wet and dry paper. Then, using a bowl of warm water with a little washing-up liquid to help suppress scumming, scrub the movement in either an up-and-down action, or by scrubbing it in all directions as you might clean a saucepan.

Having scrubbed all the edges and facings, wash the plate in hot clean water with washing-up liquid and rinse under a flow of hot clean water. Dry the plates straightaway, taking care to use the puffer to blow water from the holes in the plate. It works well to use disposable absorbent paper to dry the movement parts once they are finally finished and cleaned.

If you wish to bring the plates to a more polished finish, the procedure would be as follows: ideally start with nothing coarser than 240 grit paper and then only if there are deep marking lines to remove. Otherwise, begin with 400 grade paper – 40 micron gel film – as the coarser papers make quite a deep mark in the brass, which in itself takes time to remove.

Put the clock plate in a large tray of water with enough water to cover over the plate with about 5mm of water. Secure the gel sheet round a wooden block to help hold it flat, and draw it backwards and forwards on the surface of the plate. (Do not scrub the plate from side to side unless you require a particularly coarse and rather ugly finish!) Do both sides of both plates until they have an even grain running up and down.

Place the clock plate in a clean tray of clean water.

The first side with a 400 grit finish.

When you turn the plate over, place a clean cloth in the bottom of the tray so that the clock plate does not rub on the bottom and become scratched by the debris coming from the process. Once both sides have an even finish change the water, thoroughly rinse out the cloth and replace it in the tray with clean water. The rinsing and water change are important, as failure to do this thoroughly will inevitably lead to scoring as the finish becomes finer and odd bits of the coarse finish are picked up.

Repeat the process with 600 grade paper – 30 micron gel film. The higher finish you require, the finer grade of paper/gel will be required. Work on through, using 15 gel sheet (1,200 grit) and 9 gel (1,800 grit) and so on.

A polished finish

With patience a lovely, very deep-looking mirror polish can be obtained by using the polishing compounds manufactured by Menzerna. This is a range of polishing compounds used in the automotive industry for custom car engine part finishing and also by some of the top wrist watch manufacturers. Using the different grades is described in polishing the pinions later in this chapter. The polish readily achievable with the product is demonstrated in the lower photo at the start of this chapter.

With 600 grit equivalent. Note this plate is drawn out of the water for the photo only. If you leave brass wet as in the photo, it will stain very quickly and will need refinishing.

If you choose not to use the Menzerna finishing product, finish with a soft damp cloth and Autosol cream polish. Be sure to use a damp cloth. Wash the polish off with hot water and washing-up liquid and dry using disposable paper roll. For the polishing to be worthwhile, both plates will need to be finished to the same standard. Use a sharpened bamboo kebab stick to push the polish from all the holes in the plate. This matters, as any residue will contaminate the oil in due course, and cause early wear in the movement. Re-sharpen the stick and repeat cleaning the holes from both sides of the plates until the stick shows absolutely no discolouration from any residue.

The author prefers to polish the plate edges and have a fine straight-grained satin finish on the 'face' of the plates. This adds a pleasing contrast between the polished motion-work wheels and hour-wheel bridge.

The steel arbors will have been finished on the lathe at the point they were turned and so should require no further finishing. You may wish to polish the pinion leaves; this is a time-consuming job, but does pay dividends. The movement will look better finished, and the mesh between the wheel teeth and pinion leaf will be that little bit smoother.

Traditionally this is done with diamantine powder mixed with oil to form a paste. Diamantine is still available as a fine grade for the final polish, although there are other products that work very well. For this publication the polishing compounds from Menzerna – which are used by goldsmiths and some of the top Swiss watch houses – are used. It is possible to buy a sample block about 30mm thick of each grade for a very reasonable cost. At the time of going to press with this publication, the sample blocks shown cost less than £2.00 each.

Support the pinion in a small wooden stand, hard wood ideally, although the ply shown worked well and made it easier to cut a groove to accommodate the pinion. Note the area of the block under the pinion is cut away; this is so pressure can be used to polish the pinion without the possibility of fracturing a pinion leaf (*see* photo on the following page).

To polish the pinion leaves, begin by applying either the coarse or medium compound mixed with white spirit to form a paste with a creamy consistency. Coat the end of the wooden spatula with a generous amount, and apply it to the pinions working the polish up and down each pinion leaf. Use some pressure to begin to polish the surface. Having done all the leaves, wash away the compound completely with white spirit. This is important, as is the use of a fresh spatula for

An image of the chain-wound version showing the contrast between the polished edge and the satin surface of the plates.

Sample block of coarser grade. Note the hole forming in the middle, where the author has used a light oil to mix the compound into a paste. The hardwood wooden spatula is roughly shaped to fit into the pinion leaves. It forms into the shape of the pinion as it is used.

the next finer grade of polish. If you begin with the coarse, use the medium next – or if beginning with the medium, use the fine. Repeat the process as above, and then move to the next grade, the last of which would be the yellow, very fine compound. The more time spent on each stage will result in a higher polish on the pinion leaves. The accompanying photos show the operation.

Any parts that can be polished in the lathe will save time, and will probably polish quite well. For example, all the steel screw heads can be brightly polished. As previously discussed, the arbors can also be polished in the lathe. It is important to cover the lathe bed with a clean cloth before beginning; this protects the lathe bed from any residue from the polishing papers or polishing compounds. Once you have finished, throw the cloth away.

The screws are polished individually by placing each one in a close-fitting collet. Bear in mind that the head of a pan-headed screw is not flat. They will polish well, but will not be flat unless you use the lapping wheel as discussed earlier in the chapter on screws. Provided you are happy with a slightly rounded head, the screws can be polished using a 'buff' stick. These are used dry rather than with water when used on the lathe.

Begin with a reasonably coarse paper of around 400 grit and then swap the sticks going through the

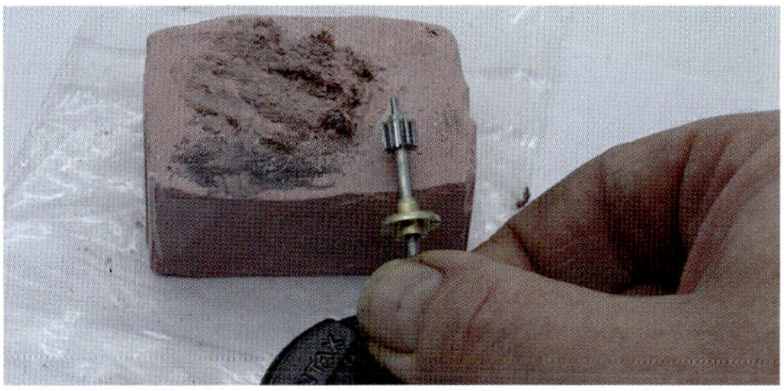

The pink compound is a mid-fine finish.

The use and shape of the block. Here the yellow compound, which is a fine compound, is being used to finally finish the pinion.

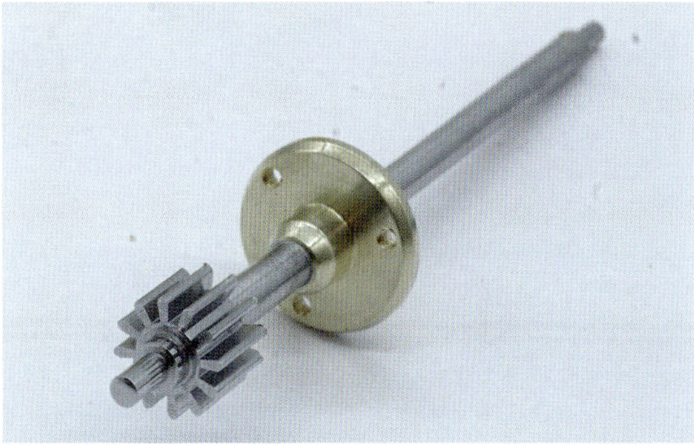

Following the yellow extra-fine compound, the pinion has a bright finish.

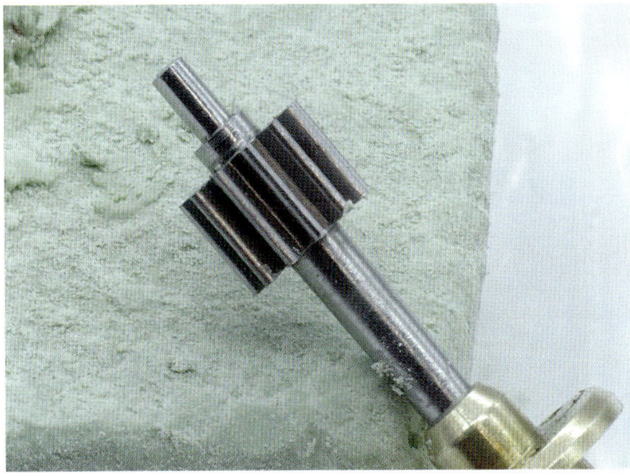

The pinion following using the coarse compound. The initial stages of finishing the pinion are done using the coarse compound, which removes the marks from the actual cutting of the pinion, but is dull in appearance and not yet finished well enough.

different grades until 2,000 grade is used. To finally polish the screw, make sure there is no residue from the buff sticks by washing them in Stoddard solvent, and use a hardwood buff stick with no paper attached, coating it with Autosol cream polish.

Once all the screws have been polished, wash them all in white spirit to remove any residue. As stated, this technique of polishing will not produce a 'flat' top to the screw head, but a slightly rounded look. The larger-diameter heads on the main plate screws do look better when the heads are flat, so if you have one, use the lapping wheel for these.

If you are using a three-jaw chuck to hold the screws, a useful way to hold the screw is to make a tube to mount in the chuck. Use a piece of brass bar of outside diameter approximately 12mm, and drill a hole through the centre to be the diameter for tapping the thread of the screws you wish to polish. Each individual screw can then be screwed into the tube and held, rather than clamping the thread of the screw in the chuck and damaging the threads.

17 The Weight and Pulley

THE WEIGHT

The weight provides the driving force for the clock. The better the clock is made, the less weight will be required for it to operate successfully. The actual weight required to drive the clock movement needs to be established. The clock photographed for this project has a traditional lead weight of 3.3kg.

Depending on the space and location of the clock the weight can be made of almost anything that is reasonably dense. It will need to be thin enough not to foul the pendulum as it passes. If using the key-wound barrel the weight will need to hang onto the pulley in such a way as the pulley will be vertical. A fairly dense material is usually used as a weight to allow for a small physical size; this is why lead was and still is often used. Lead has the advantage that if you would like a particular shape and you have the necessary equipment, you could cast it in a home-made mould. Please do *not* attempt smelting lead if you are unfamiliar with the process. It has a major disadvantage in that the fumes from smelting lead are very toxic! The oxygen/propane mix used earlier would be more than hot enough to melt the lead, but a physically larger flame works better. The author has used propane, without the addition of oxygen, with a 26mm-diameter head blowlamp to smelt lead for traditional weights in a very well-ventilated environment.

Another approach is to use a chromed or brass shell for the weight. These look very smart and are readily available. Using a shell allows for an easy adjustment of the weight as it can be filled with lead shot, sourced from a local gunsmith. The shot does not require smelting – you simply remove the brass top cap of the weight shell and pour in the shot. Measure the weight by using the kitchen or post scales.

The weight hanging from the chain-wound barrel. As can be seen, no pulley is used in this arrangement.

Around 3.3kg will be sufficient to run this clock movement. You may need to adjust the weight, but try the clock with 3.3kg and see firstly whether it continues to operate, and if so, note the angle of swing the pendulum develops. For both escapements used in this design the angle of around 4 degrees would be suitable. If the swing is very vigorous remove some of the weight and re-test. If the swing – action – of the escapement is poor you may need to add extra weight. Take care though, a poorly made movement and – or – an incorrectly adjusted escapement will not perform well however much weight is applied. As a guide, if more than approximately 4kg of weight is required you have a problem that requires attention and not more weight. Note steel shot is not heavy enough.

The key-wound barrel arrangement for the weight; a pulley is required.

THE PULLEY

This can be a plain brass pulley wheel or crossed to match the train wheels. There are commercially available pulley wheels should you not wish to make your own.

The pulley assembly consists of three parts: the pulley wheel; the eye for the weight hook; and the framework to support both pulley and weight. The pulley blank is made in the same way as a wheel blank, being cut from a sheet of brass, but instead of cutting teeth around the edge, a groove is turned into the edge of the blank to locate the line supporting the weight. Before cutting the blank out, mark-out the crossing-out lines, and the inner and outer diameter lines if crossing with six crossings.

Use the mandrel of diameter 8mm to mount the blank in the lathe, and turn the diameter to 50mm. This matches the diameter of the barrel, meaning the line coming to the pulley and heading back up towards the movement will be parallel.

Next, the groove for the line to engage with the pulley is turned into the edge of the pulley. This can be done by either using a pointed cutter angled to make a 'V' in the rim, or by using a round-end tool, turning a semi-circle into the rim. This is neater, and matches the shape of the line that will fit into it. If using a pointed tool you may use a needle file to round-out the bottom of the groove. Take care, as the file will quickly clog and could grab, spoiling the work. Also be careful not to touch the outer edges of the groove, as a file mark across the edge is unsightly and will need to be turned away.

You now have a pulley marked-out for crossing and with a groove in the edge. It would work well without crossing, but if it is going to be visible it looks neater if crossed. Use the same pattern as chosen for the wheels.

Remove it from the lathe mandrel; drill and cut in the same way as a wheel blank. Take care not to catch the edge of the line groove when handling the blank.

The author made the pulley from 5mm-thick brass. This might appear to be a bit 'chunky', for want of a

Two traditional pulleys, from the eighteenth and early twentieth centuries.

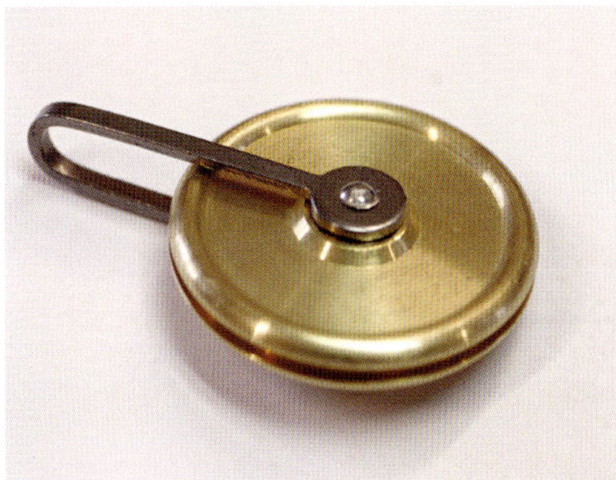

A current pulley available from the horological supplier.

The pulley – made for this project.

The groove cut into the edge of the wheel.

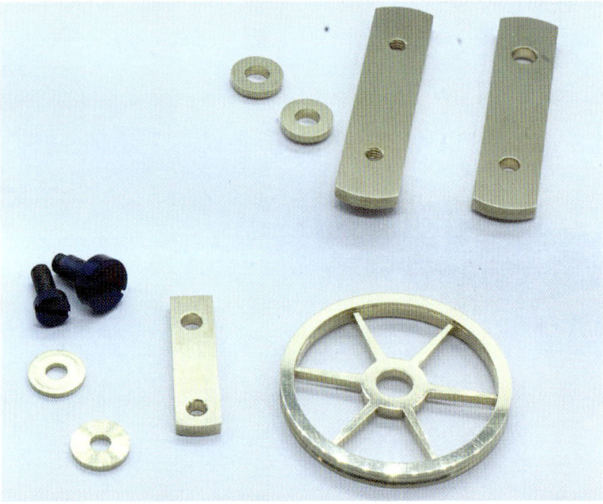

The parts required for the pulley. The dimensions are in the accompanying drawings.

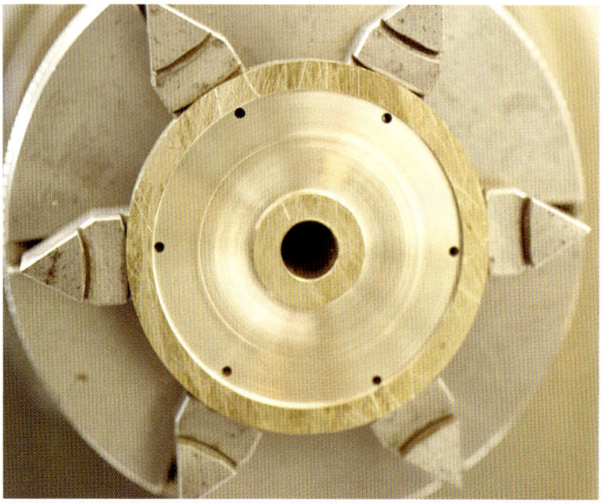

The pulley held in the six-jaw chuck.

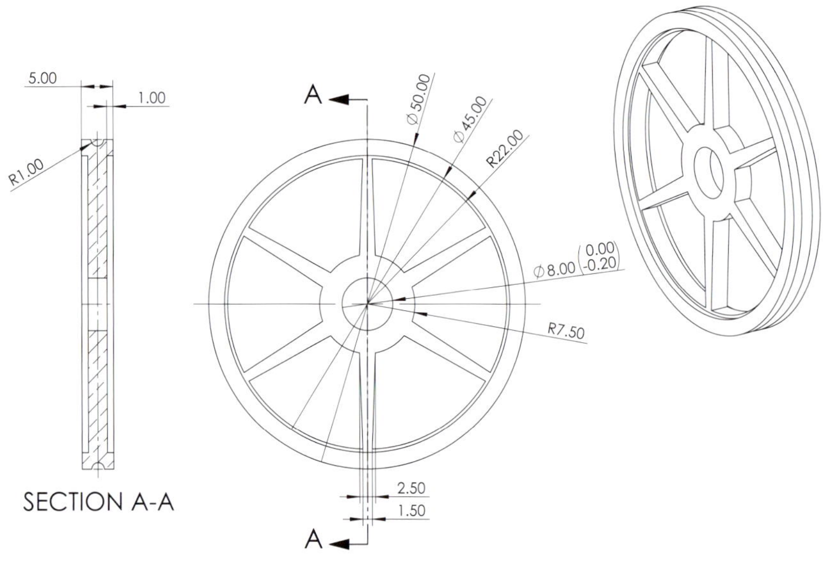

Pulley wheel.

Use the gel sheets cut into strips wrapped over the end of a flat needle file.

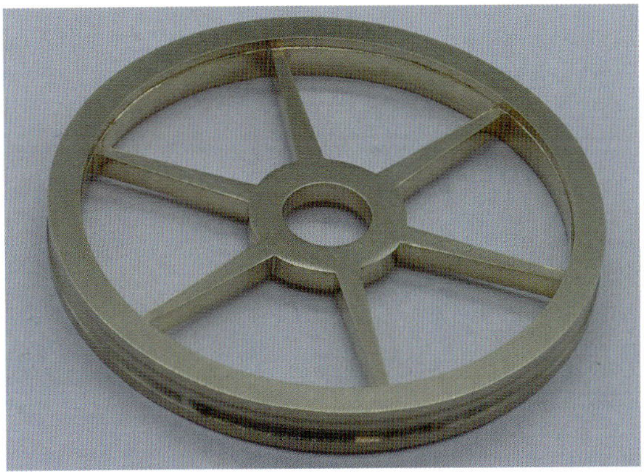

The crossings after working through to a fine grade of gel sheet.

better word. The author wished to reduce the thickness of the central section of the pulley to give it a more elegant appearance. This is an operation for which the author finds a six-jaw, self-centring chuck helpful. Bear in mind though that using a thinner piece of brass is more straightforward if you do not wish to reduce the thickness of the central section.

Reducing the thickness of the central section of the pulley

This can also help with the time spent sawing the crossings, as the thickness to cut through can be reduced by 1mm. Place the pulley in the chuck and turn to the dimensions shown on the side you have *not* marked the crossing out on.

Next, remove from the lathe and cut the crossings, taking care not to cut beyond the line marked for the outer diameter for the crossing. This diameter is slightly less than the diameter turned to reduce the thickness of the wheel. As the crossing is primarily for decoration in the case of the pulley, turning the diameter in this way will give the visual effect of a matt outer circle and polished inner, as you can see in the previous photos.

Having crossed the pulley, return it to the lathe with the side used for marking-out shown ready to be turned. Reduce the thickness of the inner section by 1mm on this remaining side. The inner section is now 3mm thick and any remaining marking-out lines have been turned away as an added bonus.

It is a slightly more time-consuming operation to finish the sides of the crossings, as they are recessed below the rim of the pulley. Care taken to make fine cuts when reducing the thickness helps reduce the time to finally finish the pulley sides.

A further polish can be achieved by using the Menzerna polishing products or similar as described in Chapter 16. For polishing the pulley, a cloth is used rather than a wooden spatula.

The roughly cut and drilled side frames. The plastic cover is as the brass was supplied. It protects the surface of the brass, making it easier to finish.

The pulley then needs a bearing surface to allow it to rotate smoothly inside the frame supporting the weight. Traditionally just the hole in the pulley and the steel of the centre screw make the bearing for the pulley. It would also be possible to fit a ball-race bearing inside the pulley.

For this pulley the bearing is made by turning two stepped washers in the same way as the movement plate washers were made. The drawing shows the dimensions.

Make one washer for each side of the pulley, resulting in the effective thickness of the central part of the pulley being 6.5mm. This gives 1mm clearance between the pulley and frame.

Commercially supplied pulleys have a 'U'-shaped steel cradle for either the weight to hook into, or for a steel hook made to join the eye of a weight to the cradle of the pulley.

Here we shall make a feature of the pulley and its frame by making the frame parts from 3mm-thick brass, and securing them together with two steel-blued screws.

The side frames of the assembly are made from two rectangular pieces of 3mm-thick brass, and can be shaped to suit the overall style of your clock. Shown are the roughly cut and drilled side frames. The plastic cover is as the brass was supplied. It protects the surface of the brass, making it easier to finish. For decoration the side frames can be pierced. The author tends to pierce the front frame. Mark out the brass, using the same techniques as previously used, to the dimensions shown.

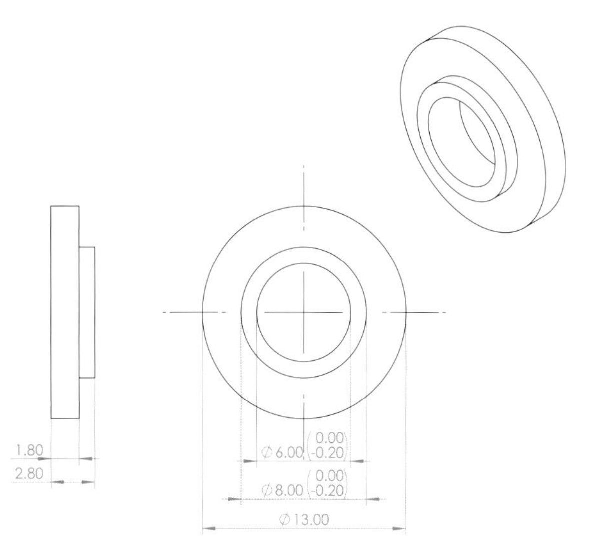

Pulley spacer.

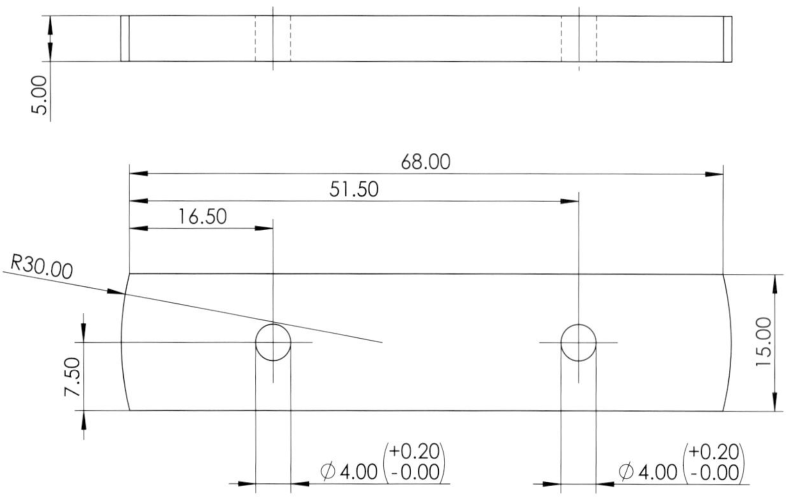

Pulley side plate.

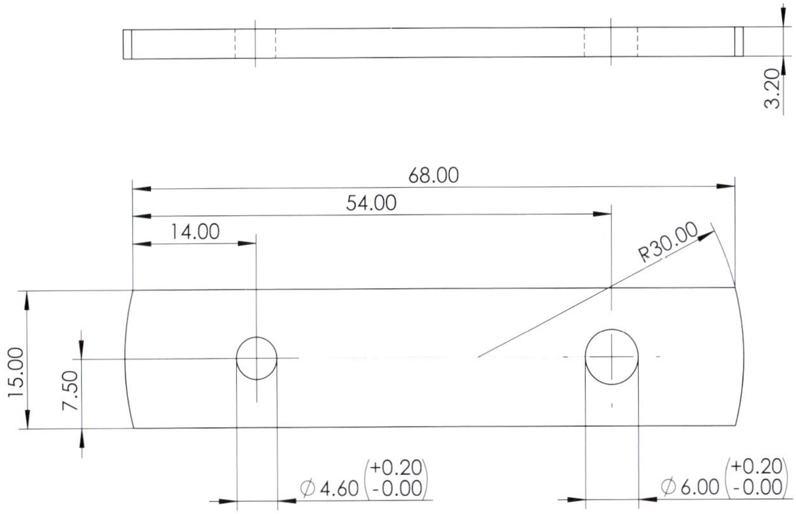

Pulley side plate 2.

Centre-punch the two holes to be drilled in one frame, and clamp the two pieces together on the table of the pillar drill. Drill both holes 3mm diameter in order to use a 4BA tap to form the thread. Next, decide which of the frames will be the front and which will be the back. Drill the holes in the front frame for a clearance for a 4BA tap diameter 3.7mm.

In the photo the shouldered screw is being used to locate the larger-diameter hole whilst the sides are filed to shape. Screw the two sides of the frame together and finish the edges so they are a matching pair. The shouldered screw is made in the same way as the screw for the click in the barrel assembly.

To hold the frames apart, and to allow the pulley to run vertically between them, a spacer is needed for the bottom screw. There is also the need to provide the hanging point and two thinner spacers for the weight to be suspended from. The washers are made in the lathe and the hanging arm from a 3mm brass sheet. You will now be familiar with the procedures required to make these parts. The dimensions are shown on the accompanying drawing. Once all the parts are made, finish them to match the movement.

The shouldered screw is used to locate the larger diameter hole whilst the sides are filed to shape.

The shouldered screw with the thread end tidied to a dome shape.

The assembled pulley arrangement.

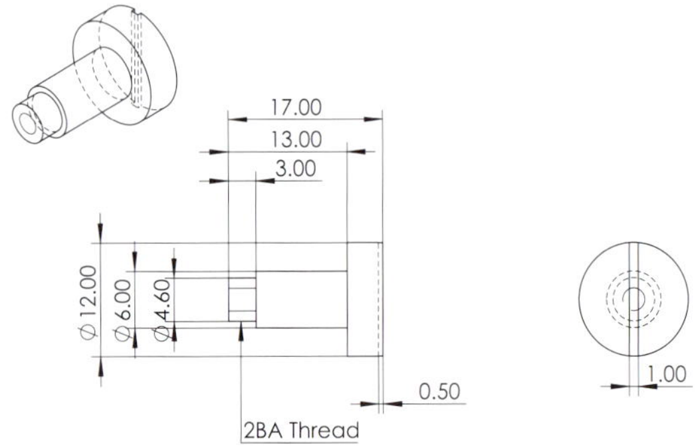

Shouldered pulley screw.

THE STAND

The seating board

Made from either a piece of solid wood, such as beech, pine, mahogany, or a section of 18mm thick good quality plywood, the seating board is designed to 'seat' the movement in its case or stand. There are many good cabinet maker's tools that can be used for shaping the seating board. Here the author has used what you might have in the general workshop toolbox.

Two designs for the seating board are shown: one for the key-wound barrel and one for the chain-wound barrel.

The key-wound clock

Mark out the seating board on what will be the underside, so if using solid timber choose the least attractive side. Follow the dimensions shown in the drawing. Use a hole saw of 16mm diameter in a cordless drill or the drill press, and cut two holes to form the slot where the weight line will pass through.

Use a chisel or jig-saw to cut away the mid points generated by the hole saw. Next use a drill 5mm in diameter to drill the two holes for the movement

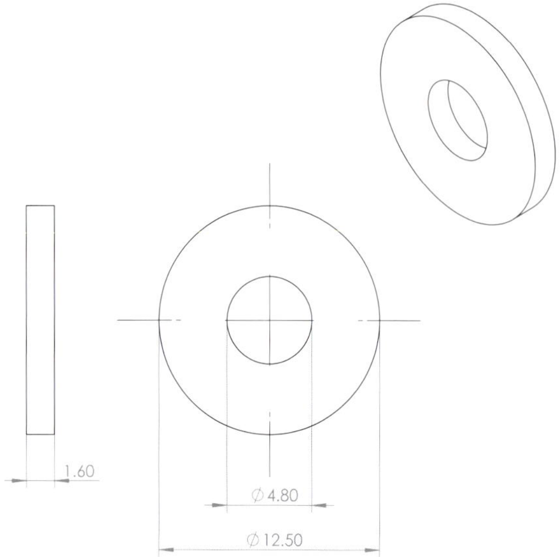

Pulley spacing washer.

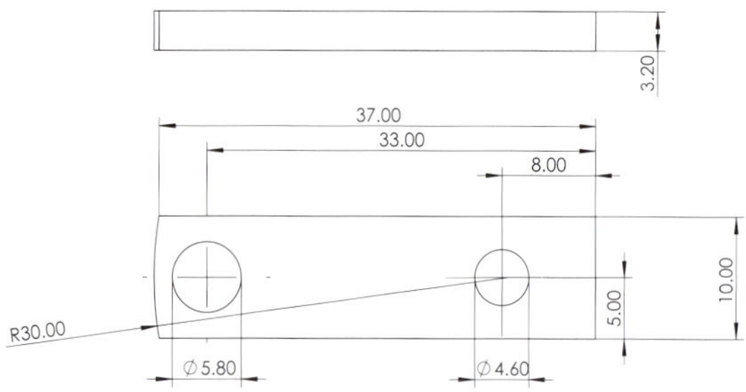

Weight hanger.

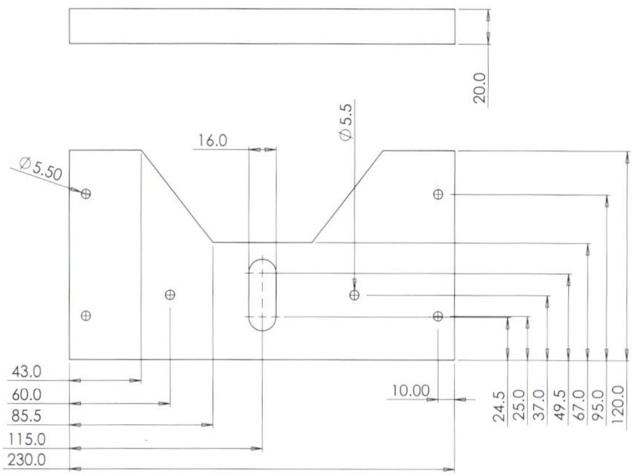

Seating board for key-wound movement.

Marking-out lines are drawn on the underside of the seating board.

Using a Forstner bit to cut the holes at the beginning and end of the slot for the line to pass through in due course. A hole saw also works well for this task.

The two holes drilled, ready for the slot to be formed by cutting away the timber between the two holes.

The finished key-wound movement seating board.

The finished chain-wound movement seating board.

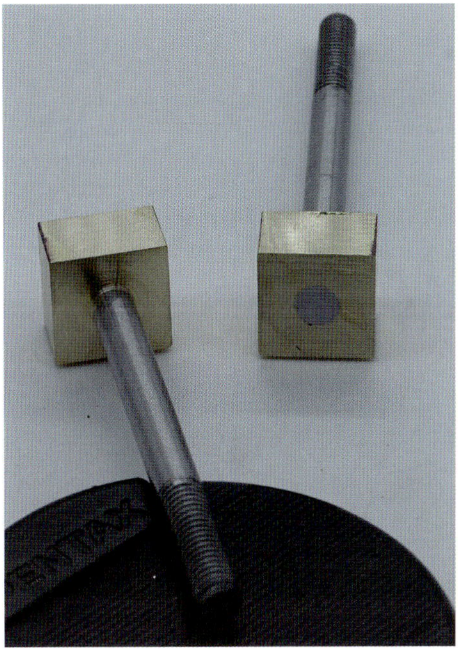

The finished mounting bolts.

securing screws to pass through. Four further holes are drilled, two at each end to allow the board to be screwed to the test horse, or into the clock case. The final hole to drill at this point is for the tail end of the line to be secured through once the movement is finally assembled. All the hole locations are shown in the drawing.

The chain-wound clock

For the chain-wound board two holes 16mm in diameter are used, rather than forming a slot. Having drilled the holes and formed it, the board can now be sanded with 400 grit sandpaper and a coat of wax polish applied.

The photo above shows the movement sitting on its board on a workshop test stand. Note the seating board is narrower than the workshop stand in this instance. Two fill-in pieces of timber are screwed in place to make a secure mounting. The actual length of the seating board is not critical. It can be lengthened if required.

Movement securing bolts

Having made the seating board, two bolts are required to secure the movement. These consist of a steel rod 4.8mm in diameter and 50mm long, with the top 24mm threaded with a 2BA die in order for them to screw into the mounting holes previously formed in the lower movement pillars. A brass square 8mm thick and of the square across the flats measuring 12.5mm is silver soldered to the other end of the steel. The actual size of the square is not important, so if you have some square bar roughly this size do use it instead.

Place the bar in the lathe – a collet is useful here as it will hold the square – and drill a hole 4.8mm in diameter through the centre. Part off a section 8mm long and repeat for the remaining screw. Silver solder this into place, remove the flux with boiling water and return to the lathe. The brass head can now be turned to true and both sides tidied. Remove the fire skin on the steel section using an old file with micro finishing film attached. Use this to also tidy the flat sides of the square.

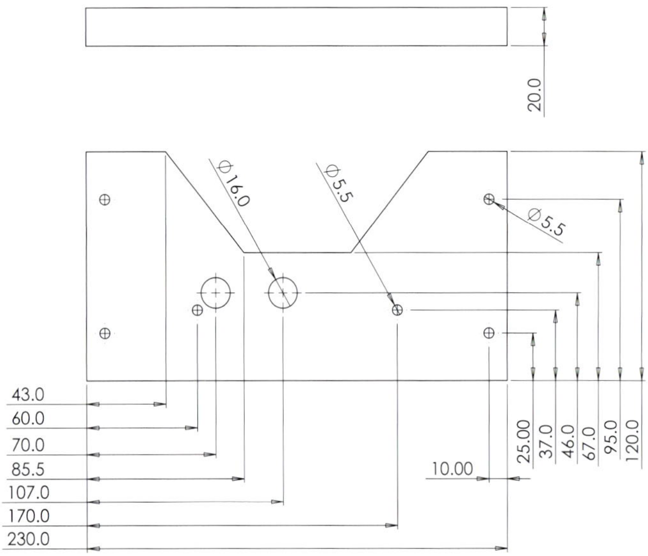

Seating board for chain-wound movement.

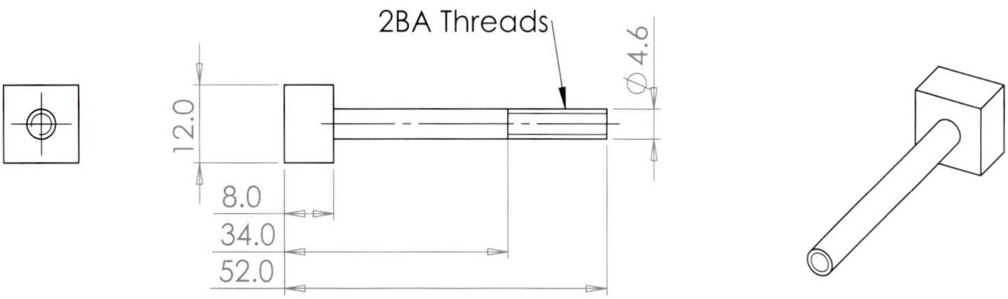

Mounting bolts.

Telling the Time and Completing the Movement

18

MAKING THE HANDS AND DIAL GUIDELINES

In the end the function of the clock is to tell the time. For this movement it is done in the conventional way of using two hands, one to indicate the hour, and one for the minutes. An extra hand can be added to indicate the seconds.

Here is a good opportunity to really personalise your clock. The design of the dial and hands can be unique to your clock.

The clock must be able to drive the hands, so lightness or a counterweight for the minute hand if over 160mm long would be advisable. A typical domestic clock minute hand is normally less than 160mm, as minute rings, known as part of the chapter ring on a clock dial, rarely exceed 300mm (12in) in diameter.

Whilst the dial design is left to you, the maker, some general points to bear in mind may be useful.

Dial guidelines

The dial is generally attached to the front plate of the movement. It could be attached to the test stand. If mounting on the movement – which the author prefers – there will need to be three or four pillars to hold the dial securely and correctly spaced to the movement. Accommodation for mounting the dial to the movement was made earlier in the project by drilling four additional holes into the movement front plate. The drawing showing the front plate in Chapter 4 gives the location of these holes. Generally clock plates – or at least the front plate – are large enough to accommodate mounting of the dial. The dial needs to be mounted far enough away from the front plate to give clearance for the motion work and the underside of the dial.

The actual dial can be made from almost any material. Traditionally, the dial plate is made from either iron or brass. The author has used chemically etched stainless steel, glass and traditionally painted dial plates.

As the project in hand is a modern weight-driven clock, if you use a traditional material such as brass, you may wish to cut it out as a modern skeleton-type dial. However, a painted scene would work well if you enjoy painting. The dial is the 'face' of the clock, and the first thing seen when the clock is admired. This makes the style of the dial aesthetically important.

Traditionally, the dial is mounted so the middle of the hands are in the middle of the dial. The name and location of the clockmaker is generally incorporated into the dial. You need not necessarily follow these traditions of course – it's your clock!

The two clocks shown are numbered 23 and 24 as they are the next in the numbered sequence made by the author. Unless otherwise stated, all the clocks illustrated in this publication are examples of the author's work.

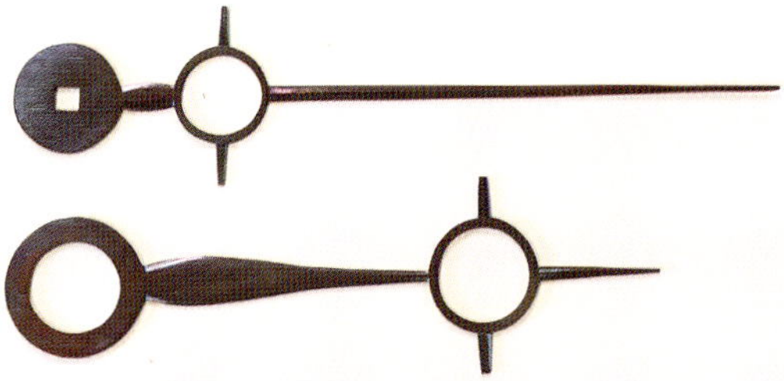

The finished hands for the clock.

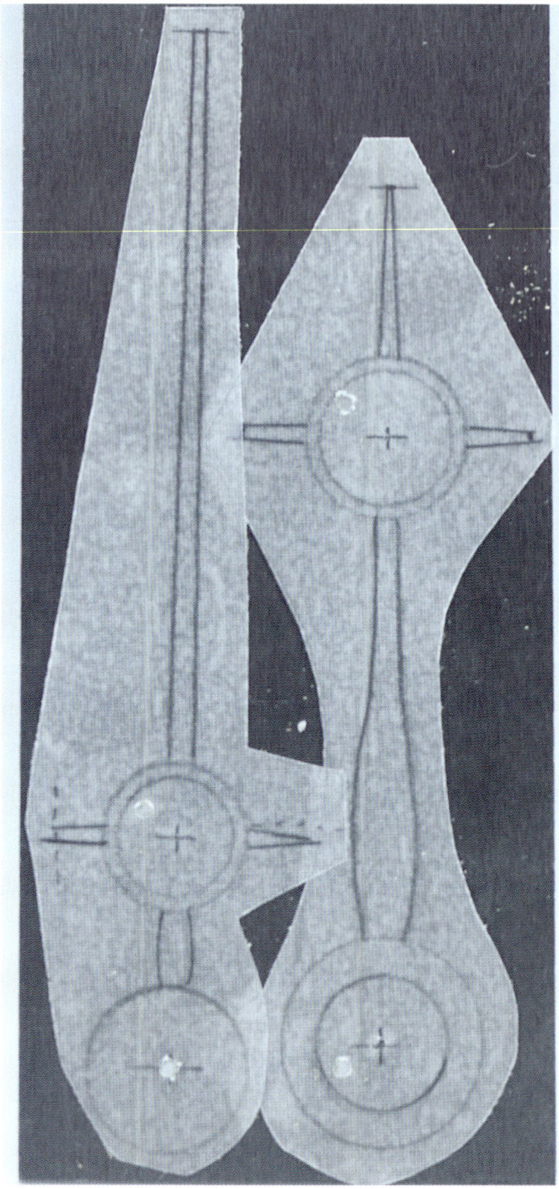

The patterns for the hands are drawn onto paper and secured to the steel using a thin, but strong, double-sided tape.

Choice of hand material

Whilst more-or-less any material can be used to make the hands, it is most usual to use brass or steel. Steel can be cut more finely than brass and still be strong enough to use. The author has used titanium for long hands, which works well, but it is hard to work with hand tools, and care must be taken when drilling. Aluminium is light, but can be bent easily and so only tends to be found on quartz battery movement powered clocks, where the light weight and low costs are of concern. Carbon would be an interesting choice also.

A good-quality steel – but not stainless steel as it is hard to work with hand tools – is a good option as it can be heat-treated to give a lovely straw/purple/blue coloured finish. Brass hands can be polished and chased to finish, perhaps gilding to add a special finishing touch. To illustrate, shown is making and finishing a pair of hands cut from 1.2mm precision ground flat-stock steel. The minute hand is 140mm long, the hour hand 90mm, and the seconds hand from centre to tip 50mm.

To begin, draw out the design of your hand design full size, this can then be stuck to the material chosen and cut out using the piercing saw. Any shapes within the hand design with a continuous edge will need a 1.2mm hole drilled into each area, to allow access to use the piercing saw. The author chooses to cut out as much of the hand as possible whilst it is part of the steel sheet it is being cut from. This gives more material to hold whilst sawing. Sawing along the lines as neatly as you can will save time on filing the edges to tidy the hand. All the edges will need to be filed, but provided your filing is neatly done, it will probably not be necessary to spend time giving a high finish to the inner edges.

The outer edges need to be finished by filing, draw-filing and using the films stuck onto a needle file to give a neat, smooth finish. Using needle files in this way gives a useful way of finely finishing the details of the hands. Bear in mind that the file will need washing in white spirit to remove the sticky residue left from the film sheet when you have finished.

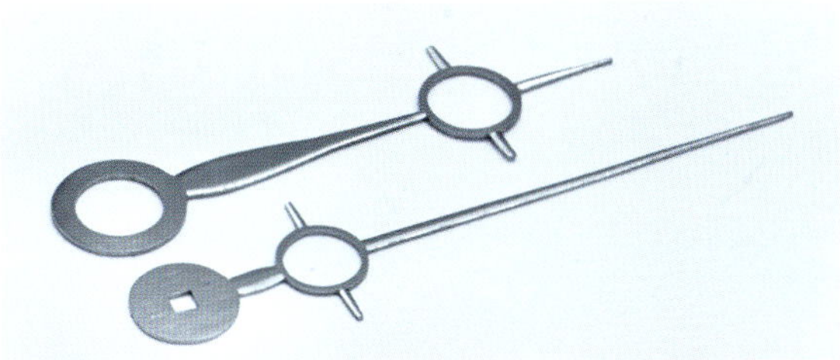

The edges of the hands catch the light when finished well.

Use wet and dry paper or film sheets to smooth the underside of the hand to make it easier to handle. The front of the hand can either be left flat or have aspects of it chased to give a three-dimensional visual effect. The author has chased the hands shown above to give an appearance of a star. Any surface chased will need to be finished.

File the centre of the minute hand so it fits firmly onto the square filed on the end of the cannon pinion tube. The centre of the hour hand is a circular hole of a diameter to allow a brass collet to be made to secure it to the hour wheel pipe.

The last surface to finish and then polish is the front surface, giving the steel the best finish you can. The better it is finished, the more likely it is to turn a deep royal blue colour when heated. If you are not bluing or plating the hands, it is not necessary to polish the front surface. A 'smoothed' surface would help a painted finish to hold more securely to the surface of the hands.

Bluing steel hands

The process of hand bluing may take several attempts to achieve a good even colour, but is worth the effort. At the beginning of the book it was suggested that any brass filings produced should be kept to one side; now is the time to use them. This process is carried out in the hearth and a well-ventilated area for safety. Do not use brass filings contaminated with white spirit as it is very flammable. Wear safety glasses.

To obtain the final blue finish with steel hands, heat is applied in a particular way to change their colour. It is worth stating once again that the hands will only blue well if the steel is of good quality and the surface is well polished.

This process is carried out once the hands have been totally finished and the front surface polished. Wash the hands in white spirit and dry using a paper towel. Take care not to touch them with your bare hands from now on. The technique used is the same as for bluing the main movement screws, although this time a copper tray is used to give enough length to fit the whole hand in. Blue the hands one at a time as it is easier to gain a constant colour change, and if it goes wrong it is just the one hand to redo!

Spread the brass filings out in a copper tray. The first time the fresh filings are heated there will probably

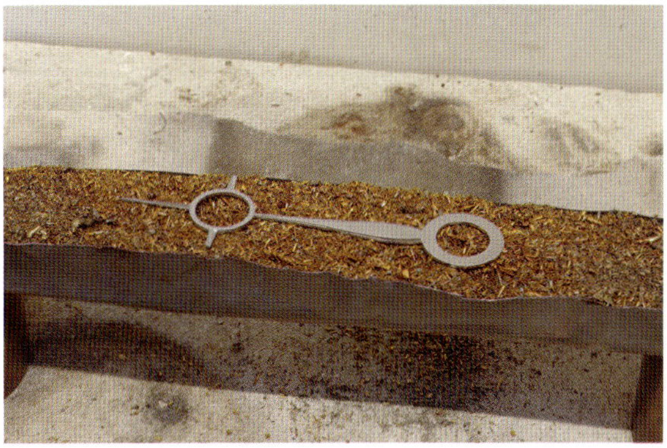

Spread the brass filings out in a copper tray.

be some smoke rising from them. This is residue from when they were filed burning off. Wait until there are no fumes of any type coming from the brass filings before leaving them to cool down ready for use. The next time the filings are used there should be no smoking or residue.

Place one of the hands onto the filings and, where the hand is thinnest, use a pair of tweezers to push the filings away from the underside of the steel to form a slight gap. Where the steel is thickest, make sure there are plenty of filings touching the steel. It is the heat transferring from the filings to the steel that will cause it to change colour. The thinner the steel, the less heat will be required. The idea is for the whole hand to go a uniform colour. In order to achieve this the thicker parts require more heat, and the thinner parts less. It takes practice to achieve a uniform finish. If the steel gets too hot it turns a rather dull grey colour. In the event of this happening, remove the hands from the filings and let them cool down – do not quench them. The surfaces will need to be re-polished before they can be placed on the brass filings for another attempt.

Note the final photo is not taken with the hand on the filings. Once the hands are the required colour they must be removed straightaway and placed on something like a flat piece of brass to help absorb the heat as quickly as possible. If they do not cool quickly enough, the colour will go over and the process will need to be repeated from the beginning.

An alternative to re-finishing the hands, and starting again if the colour goes over, would be to use an oil bath to quench the hands whilst they are hot. This will not give a royal blue colour, but will give a traditional dull black/grey/dull blue finish without the need to begin again. Heat the hands a little more so that the matt grey colour is even – with no blue

The heat is applied under the tray of filings, moving the torch along the length of the tray. The hand begins to turn a straw colour.

The heat is continued to be applied and the colour begins to change to a purple. As can be seen the hand is not changing colour in a uniform fashion.

To help even the colour change, the filings have been moved from under the hand where it has changed to a dark purple, almost blue.

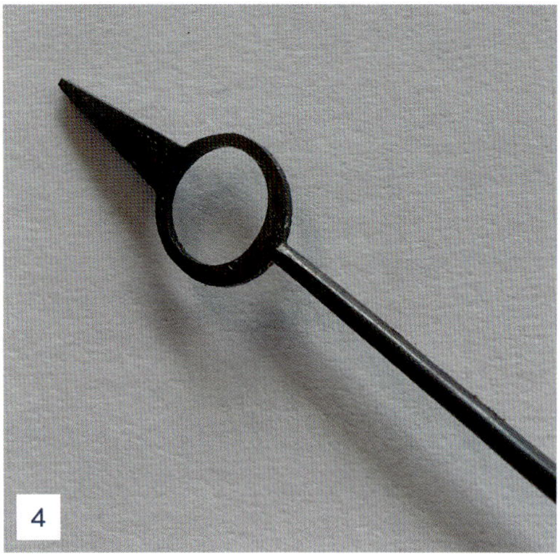

The finish once the hands have been heated to grey and plunged into oil. This gives a uniform but dull dark-grey finish. The finish is quite good at resisting rust, however.

being left – and then plunge the hands into a tray of oil straight from the brass filings. Care must be taken here; there must be plenty of oil in the tray as the hands need to cool down, and the oil must not burn. The oil will cool them and give a uniform dark finish. Plenty of ventilation is required and it is best done outside to save smoke alarms triggering and any mishap occurring.

FITTING THE HANDS

The minute hand fits straight onto the cannon pinion, and is held in place by the minute-hand washer. The cannon pinion has a square filed onto the end, to which the minute hand is a firm fit. Use a square needle file to form a square in the minute-hand hole, the diameter of which is shown in the drawing. Once you have the rough shape of a square, check often to see that only enough steel is removed to be a good fit.

A close-up photo showing the lower section of the minute hand and the fitting of the cannon pinion square end into the square hole in the hand.

A domed brass cap known as a minute cap is then fitted to hold the hand in place.

The thickness of the collet determines how compressed the friction spring under the cannon pinion will be. The cap and tension are secured using a steel pin as shown in the photo overleaf.

This friction ensures the hands travel round with the centre arbor, but also allows the time to be set.

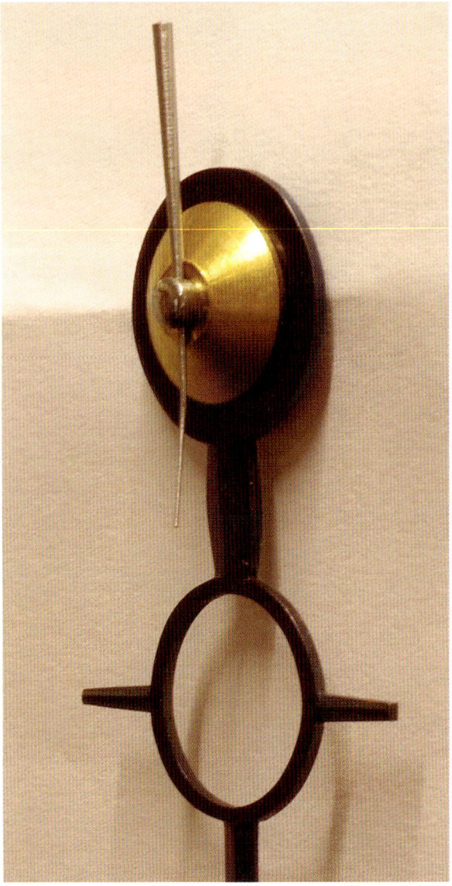

The cap and tension are secured using a steel pin.

The dimensions for the minute cap are given in the accompanying drawing. The process of turning the minute cap is begun by placing a round bar section in the lathe and turning a length of approx 20mm to the diameter shown. Using the diameter shown will mean when the cap is fitted to the minute hand there will be a blued band of the minute hand showing round the edge – this gives a good visual effect. (This can be seen in the photo.) Next, drill a hole through the centre to match the diameter of the centre arbor you have made, plus 0.1mm.

A section approximately 1mm long is next turned to a diameter of 6mm. This section has a two-fold role in giving a pleasing look, and gives a step to hold the reversed minute cap in the lathe for turning the underside. If necessary this step can be turned away later to adjust the fit and tension of the minute hand.

The final length of the minute cap depends on where the pin hole is actually drilled in the end of the centre arbor. This is shown on the drawing for the centre arbor. It is important to check where the hole you drilled is located in respect to the end of the centre arbor.

Having drilled the arbor pin hole, the length of the minute cap can be determined by placing the hand tension spring on the arbor, then the cannon pinion on the spring. Can the pin hole be seen in front of the cannon pinion? Measure the distance from the front of the cannon pinion square to the pin hole and add 0.5mm. This should give the thickness of the minute cap required. The idea being that when the minute cap is fitted, and a pin passed through the hole – *see* the photo – the tension spring should be compressed enough to hold the minute hand in place – whilst allowing the minute hand to be adjusted to correct the time.

The hour hand collet

The hour hand mounted on the collet is a friction fit to the hour wheel pipe. The thickness of the hand alone will not grip the pipe sufficiently, so a brass collar is

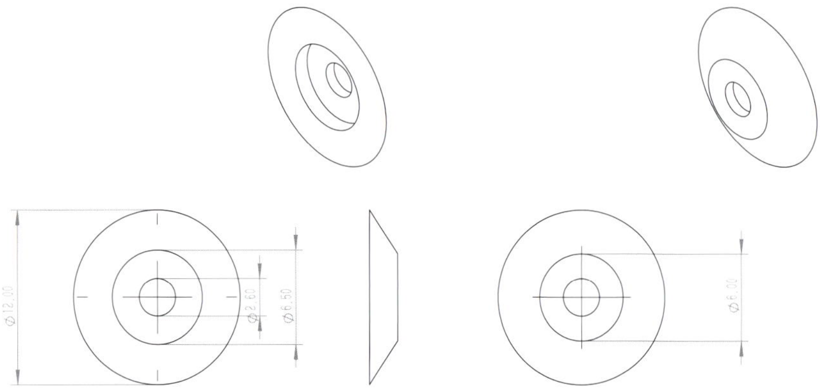

Minute cap.

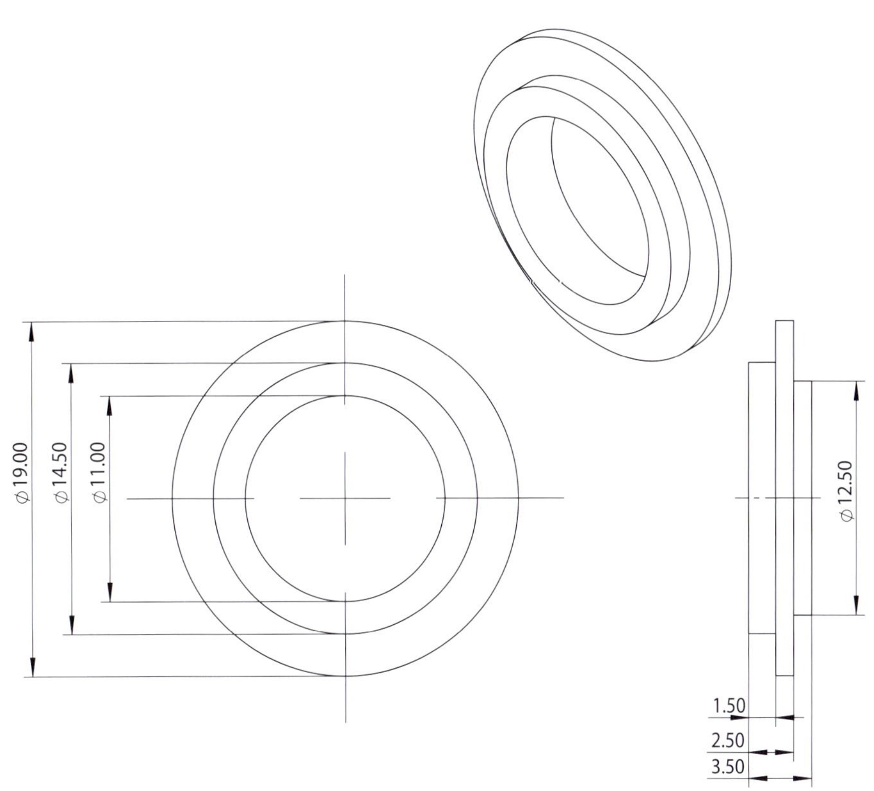

Hour-hand collet.

made and fitted. As can be seen from the drawing, it is a brass collar with a thin wall and a step for the hour hand to be either riveted to or secured using Loctite 638. Traditionally the hour hand is riveted, but the Loctite product or similar will work well and save possible damage to the hand – or the possible reduction of the inner diameter of the collet stopping it fitting correctly. The inner diameter of the collar needs to match the diameter of the hour wheel pipe, so a careful measurement of what you have actually made is required. In the case of the author's hour wheel pipe, the actual diameter is exactly 11.0mm. To match the collet, the author has turned the collet outside diameter and then drilled a 10mm diameter hole. The boring bar tool was used to turn the diameter of the hole to the final diameter.

Try the collar on the hour wheel pipe whilst it is still mounted in the lathe and make only very small cuts to come to the size required for a push fit. If you go a little too far you can use the collet by cutting down either side of the underhand section with the piercing saw, and lightly squeezing the collet. You might prefer to start again, of course.

FINAL CLEANING AND ASSEMBLY OF THE MOVEMENT

Clean, well-finished brass looks lovely when assembled as a clock movement. It is important not to put fingerprints on any of the movement as you assemble it. Finally cleaning the brass in the world of commercial horology is carried out with products developed especially for the task. These are not readily available to the general public and require the correct disposal.

To add a final finish to the movement, all polished parts can be polished using a soft bristle brush and a cream polish such as Autosol or Peak. Other polishes are available, such as the polishing compounds manufactured by Menzerna for example. With practice and plenty of time, it is possible to give a good finish to the brass. Apply a small amount to the brush and work it over the surface of the item to be polished.

Use a fresh soft brush without polish to remove the polish. Then place in a bowl of hot water with washing-up liquid – a green one works well – and use a toothbrush to remove any remaining polish. Rinse the part

The use of the boring bar for this operation.

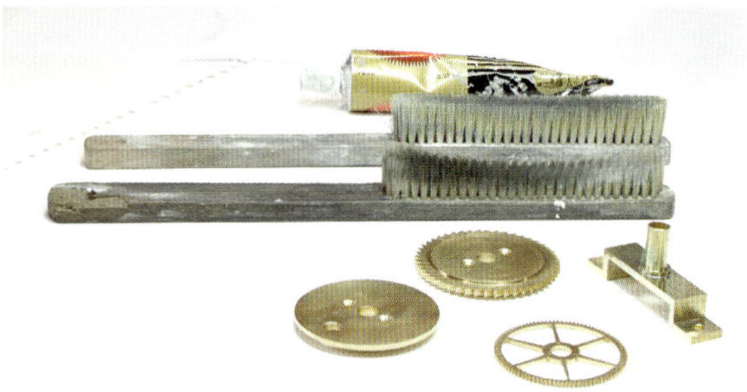

Brushes and polish ready to add a final finish to the movement parts.

under clean, hot water and dry immediately with a paper towel. The result will be a clean and polished part. If you are polishing a flat polished area such as one of the clock plates, use cotton wool. For polished edges a paper towel with polish added works well.

Once the parts have been polished and finished any holes need to be 'pegged out' to ensure they are clean and polish-free. This is very important, as the oil would

The cream polish applied to the brush ready for action.

be contaminated by the polish, and the movement would quickly wear. Use a sharpened bamboo barbecue stick for this; these are cheap and readily available in the supermarket.

Assembling

Having come this far, now comes the reward for all the hard work. Great care must be taken when handling the finished clock parts as any fingerprints will cause steel to rust, and brass to discolour. It is best to use gloves when handling the movement parts – purple gloves made from Nitrile are readily available over the internet. (Although the colour or material does not actually matter!)

By now you will be familiar with how the clock movement fits together, and so it is now time to assemble. Begin by assembling the constituent parts. The barrel needs the line threading through the hole in the barrel shell and the end tied to stop it pulling back through the barrel. If fitting a gut line, the end beyond the knot is melted using a match or lighter to stop fraying. For bronze and brass lines the end is soft soldered using a spirit lamp to heat it to stop it unwinding. Do not use steel lines – they belong to bicycles and cut into the surface of the brass barrel.

Assemble the rest of the barrel, greasing the points that rotate against each other as shown in the photos.

Carefully wrap the line round the barrel ensuring each turn touches the previous one, but does not cross over. Once the barrel is full, secure the line round it with masking tape to stop it unwinding. Measure a further 400mm of line and, if using brass or bronze line, melt some solder at this point about 25mm long. Cut the line

The assembled, cleaned and finished barrel ready to be greased before the great wheel is fitted. The grease used in this instance is Mobius 8300.

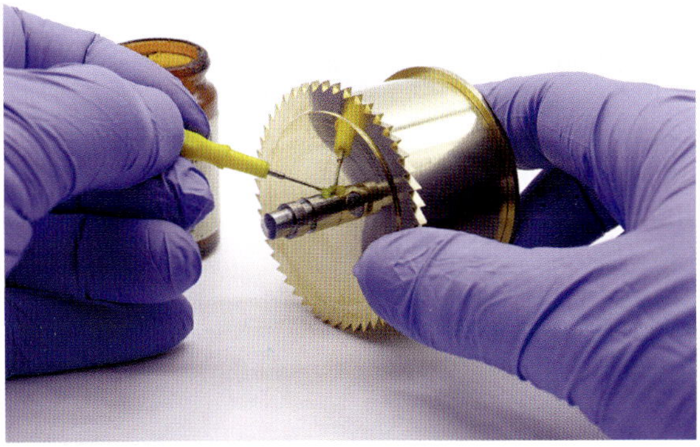

Handling the movement with gloves for assembling avoids fingerprints spoiling your work.

Adding grease to the chain-wound barrel assembly.

off halfway along the solder; neither end of the line will untwist as the solder holds it in place. For the gut line, cut it off and scorch the end with a match or lighter.

The main movement pillars are then fitted into the backplate and secured with their relevant finished screws and washers. Place the movement on a soft cloth or movement stand on its back plate and place the barrel assembly, wheels and maintaining power iron if fitting one, into their pivot holes.

Take great care not to catch the end of the arbors at this point, as this could well cause a bent pivot. Place the front plate onto the movement, locating the centre arbor, the barrel arbor and then escape wheel arbor as the plate comes down. Finally the third wheel will fit into its pivot hole. Secure the front pillar locating screws to make the movement safe.

The movement parts assembled and located ready to fit the top plate.

Next, the front centre arbor pivot is lubricated with Mobius D4 clock oil and the friction spring fitted. Grease the facing tongues of the friction spring, and the forward section of the centre arbor before sliding the cannon pinion into position.

The hour-wheel bridge fits next, screwing it into place. The hour-wheel assembly fits onto this, but without lubrication. Double-check there is no polish whatsoever inside the hour-wheel pipe. Oil the reverse minute wheel post and place the reverse minute wheel onto its post. Fit the securing nut. This completes the front plate fittings and so the movement can be stood on its base. All should be looking smart and well at this point. At this time it is useful to mount the movement on the seating board, threading the line through the slot. This gives a way of holding the movement without having to touch it.

Next the pallets are fitted into place and the pallet bridge fitted to secure them.

The complete movement. The chain-wound version with the recoil escapement is shown here.

Lubrication

It is important to use the correct lubricants for any clock movement. Recommended is the Mobius range of lubricants; for this clock D3 and D4 are ideal, although D4 would suffice, along with Mobius 8025 grease. These are available from the suppliers listed in the Appendix.

Oil the escape- and third-wheel pivots with D3 and centre and barrel pivots with D4. The pallet impulse surfaces are lubricated with D4, as this will stay in place better than a thinner oil. Next, fit the two hands and secure with the brass minute cap. Set them at 12 o'clock to check that they are aligned correctly.

Place the movement secured to the seating board on the test stand and fit the pulley. Pass the line through the pulley and up through the hole drilled for this purpose. Tie a knot large enough to stop it coming back through the hole. If using the chain-wound barrel, fit the chain at this point.

Next fit the pendulum, and hang the weight on the pulley. Swing the pendulum and the clock should tick – **well done!**

ADJUSTING THE BEAT

For the traditional crutch, bend the arm to the left or the right. You will notice the beat change. Continue adjusting until the beat is even.

For the mechanical adjuster, loosen one of the adjusting screws and then turn the other screw to move the crutch arm. Continue until the beat is even, making sure both screws are tight once finished.

POSTSCRIPT...

To illustrate this publication two movements were made, both of which the author used to make complete clocks. These are shown at the beginning of the chapter.

The photo shows the movement fitted with the deadbeat escapement and with a key-wound barrel with maintaining power. The train was mounted in a set of circular plates and a year-calendar arrangement was fitted to the front plate. To this has been added a partially skeletonised dial allowing the calendar wheel work and the action of the escapement to be visible.

The same movement showing the rear plate and beat adjuster version of the crutch fitted to the same movement.

View of the partially skeletonised dial and front plate of the movement.

The two photos show the movement fitted with the deadbeat escapement and with a key-wound barrel with maintaining power. The train was mounted in a set of circular plates and a year-calendar arrangement was fitted to the front plate. To this has been added a partially skeletonised dial allowing the calendar wheel work and the action of the escapement to be visible.

The completed movement and dial has been fitted into a wall clock case, made from English walnut. The weight is a chrome shell filled with smelted lead shot to give a compact weight, which combined with a 34mm-diameter barrel allows the clock to run for a full week within the wall clock case. The pendulum is the basic dimensions shown in the drawings. To enhance the visual impact two chrome tubes have been used as the bob, rather than one single cylindrical bob.

The movement fitted with the recoil escapement has also been fitted with a key-wound barrel. This time the one shown in the drawings of the publication, with maintaining power.

The dial is painted so it is rather more traditional. As can be seen, a second hand is fitted. This is achieved by extending the length of the escape wheel arbor through the front plate of the clock as described in the text.

The complete movement and dial assembly has also been fitted into a wall clock case made from English oak. The current weight is the lead item used to test the movements. It will in due course be fitted with a brass shell. The pendulum is made to the dimensions shown in the drawings, the brass bob is as described. For visual effect the pendulum rod is quartz rather than steel, with a steel thread being secured to the bottom of the rod and a brass top section with a slot cut for the suspension spring fitted to the top of the rod.

This clock is shown in its final version on the rear cover of this book. For visual effect, the driving weight is divided in two, and set to run down either side of the case.

Appendix

BA TAPPING THREADS AND DRILLS

TAP SIZE	TAPPING HOLE DIAMETER	CLEARANCE HOLE DIAMETER	DIAMETER TO USE A DIE
2BA	4.0mm	4.8mm	4.7mm
4BA	3.0mm	3.7mm	3.6mm
5BA	2.6mm	3.3mm	3.2mm
6BA	2.3mm	2.8mm	2.8mm
8BA	1.8mm	2.3mm	2.2mm
10BA	1.4mm	1.8mm	1.7mm
12BA	1.05mm	1.3mm	Not required.

METRIC TAPPING THREADS AND DRILLS

THREAD SIZE	TAPPING HOLE DIAMETER	CLEARANCE HOLE DIAMETER	DIAMETER TO USE A DIE
M5	4.2mm	5.1mm	5mm
M4	3.3mm	4.1mm	4mm
M3	2.5mm	3.1mm	3mm
M2.5	2.05mm	2.6mm	2.5mm
M2	1.6mm	2.05mm	2mm
M1.6	1.25mm	1.65mm	1.6mm

SUPPLIERS

Wheels, pinions, clock keys, lines and general horological supplies

John Wardle Horological Materials
Unit 6 Albany Court Blenheim Road
Airfield Industrial Estate
Ashbourne
Derbyshire
DE6 1HA

01335 347154
sales@johnwardle.co.uk

Wheel and pinion cutters

P.P. Thornton (Successors)
The Old Dairy
Prescote Manor Farm
Cropedy
OX17 1PF

01295 279060

Hand tools

For the gel sheet starter set, including the late glass base:

Workshop Heaven Ltd
Unit 5, Alkerton Oaks Business Park
Stratford Road
Upton
Banbury
OX15 6EP

01295 678941
sales@workshopheaven.com

Eternal Tools
Sales Office
Union Square
Wootton
Oxfordshire
OX20 1DU

01993 812631

A.G. Thomas Bradford Ltd
Tompion House
Heaton Road
Bradford
BD8 8RB

01274 497171
info@agthomas.co.uk

Materials

Brass and steel suitable for this project:

Ian T. Cobb

0116 2676063
enquiries@clockmaking-brass.co.uk

Ground steel and so on:

MSC Direct
7 Pacific Avenue
Wednesbury
WS10 7WP

0800 663355
sales@mscdirect.co.uk

Brass:

Smiths Metals Stratton Business Park
London Road
Biggleswade
Bedfordshire
SG18 8QB

0845 5273331
info@smithmetal.com

BASIC MATERIALS LIST

Please note size is approximate. Note has been taken of readily available sizes when purchased in small quantities.

CZ120 brass sheet for clock plates: 3–4mm thick, 150mm × 200mm.
Three of these will give enough for the clock plates and other parts.

CZ120 brass sheet for clock wheels (sheet thickness followed by width and length):
1.5mm; 150mm × 200mm.
2.0mm; 150mm × 150mm.
3.0mm (3.175mm from Ian Cobb); 150mm × 150mm

5.0mm (4.75mm from Ian Cobb); 150mm × 150mm will allow one greta wheel to be made, and a smaller disc cut to act as a backing plate. Ideally a piece 150mm × 200 mm would be purchased.

Brass bar

12mm diameter: in total approximately 450mm (minimum length 150mm).
15mm diameter: in total approximately 250mm (minimum length 150mm).
16–17mm diameter: 150mm length.
25mm diameter: 50mm length.
34mm diameter: 120mm length (this is the pendulum bob, so if using a traditional bob from the suppliers this is not required).

Brass square bar

12–14mm square and 90mm long (for making a tool)

Brass tube for barrel shell

50mm diameter, and 40mm long to allow the ends to be faced (minimum wall thickness 1.5mm).
Brass 90-degree angle with minimum thickness of flat section 3mm. Approximate length: 100mm.
For steel bar EN8 would work well. It may be possible to purchase what used to be called free cutting silver steel.
3mm diameter, 300mm long.
5mm diameter, 1100mm for pendulum and rating.
7mm diameter, 60mm long.
10mm diameter, 200mm long.
12mm diameter, 120mm long.
16mm diameter, 200mm long.

Blue steel is purchased from the horological suppliers listed. The sizes are noted in the text.

Ground flat stock steel

Thickness: 2mm; 60mm × 30mm. A 50mm × 50mm piece should be included if making the deadbeat pallets.
Thickness: 4mm; 60mm × 30mm for the recoil pallets.

No12 Longcase crank key (John Wardle manufacture these).

FURTHER READING

Barnes, Nigel and Ilmonen, Karoliina, Clock Cases – *A Practical Guide to their Construction, Restoration and Conservation* (Crowood)

Gazeley, W.J., *Clock and Watch Escapements* (Robert Hale)

Matthys, Robert J., *Accurate Clock Pendulums* (Oxford)

Wild, J. Malcolm, *Wheel and Pinion Cutting in Horology: A historical and practical guide* (Crowood)

Index

assembling, clock 183–186

barrel assembly
- arbor 111–115
- barrel arbor 99
- brass barrel 100–105
- chain-wound barrel 95, 111
- circlip 111
- click spring 108–109
- key-wound barrel 95–99
- securing collar 115–117
- winding ratchet 106–108

bench-top anvil 28–29
bench vice 32–33
brass 19
British Standard Time (BST) 8
Brocot deadbeat design 17
Brocot pin-pallet deadbeat escapement 146
Burr lubricant wax 45

centre drill 21
chain drilling 146
chain-wound barrel 15, 16
civil year 7
clock
- CZ120 8
- dials 8
- escapement 8
- history of 8
- movement 8
- wheels and pinions 8–9

clock movement
- assembling 183–186
- brass 19
- deadbeat escapement 17
- finishing process
  - gels/films 163–164
  - plates 164–165
  - polishing 165–167
  - wet and dry papers 163–164
- lubrication 186
- options 15
- recoil escapement 17
- tools required 19 (*see also* tools)
- winding 15–17

clock plate
- cutting 39
- drill the holes for pillars 39–40
- marking-out technique
  - engineer's blue 35
  - shaped plate variant 36, 38–39
- plugs 41–42
- rectangular plate 36–38
- tidying and finishing the edges 47–48
- workpiece, drill to the 43–47

clockwise 8
collets 87–88
- to the arbor, soft soldering 89
- rivet a wheel into place 94
- rivet the wheels 91
- screw the wheel onto 90
- split-staking tool 91–93

crutch 137–139
- actual beat adjustment 142–143
- for adjustable beat arrangement 141
- beat adjuster 141–142
  - to collet 143–145
- brass mounting collet 142
- with mechanical beat adjusters 140–143
- pin 142

CZ120 19

day, types of 7
deadbeat escapement 11, 12
depthing tool 31
detached escapement 12
dial guidelines 177–178
domed friction washer 144–145
Dormer twist drills 21
drill press *see* pillar drill

escapement 8, 10
- deadbeat escapement 11, 13
- grasshopper escapement 11–12
- gravity escapement 12
- recoil escapement 10–11
- verge escapement 10
- working principle 12

file button 139–140
finishing papers and films 34
five-sided broach 30–31
Fobco 'Star' pillar drill 20

grasshopper escapement 11–12
graver 26
gravity escapement 12
Gurney regulator 11, 12

hammer 28
hand, the
- fitting 181–183
- hand bluing, process of 179–181
- material, choice of 178–179

hand files 25–26
hand puffer 34
horological wheel cutters 23
hour wheel bridge 133–135, 136

jewelled pallet pads 146
jobbing drills 21

key-wound barrel 15–17

lathe
  chuck 21, 23
  collets and chucks 23
  Pultra 1770 lathe 22
  Schaublin 102 lathe 22
  Schaublin vm102 lathe 22

machined pallet bridge 135
machine vice 32
maintaining power device 17
mean day 7
measuring and marking tools 27–28
milling machine, use of 135
motion work 125–129

oil stone 21

pallet arbor 137
pallet blank cutting, using piercing saw 146–147
pallet bridge 130–133
pallet pads 146
pallets
  deadbeat pallets 151–152
  harden and temper 151
    escapement impulse pads, the 154–155
  pads 152–153
  process of making 150–151
  steel impulse pads 153–154
pendulum
  cylindrical pendulum bob 156
  drilling a hole 157–158
  finishing process 162
  knurling tool 159
  rating nut 159
  rating thread 159
  suspension spring 159–160
piercing saw 23–25
pillar drill 20
  Dormer twist drills 21
  Fobco 'Star' pillar drill 20
pillars 48–51
pillar screws
  bluing screws 56–57
  dimensions 52
  fine cutter 54
  screw-head polishing 54–55
pinions and arbor
  finishing 86
  loose pinion technique 85–86
  making 81
  tailstock arbor support 81–85
pin wheel escapement 146
pivot file/burnisher 29–30
pulley, the 169–173

recoil escapement 10–11
  pallets 147–149

Schaublin arrangement 63–64
screwdrivers 27
screws *see* pillar screws
Sexagesimal 6
sidereal day 7
silver soldering process 131–132
solar day 7
spotting drills 21
stand, the
  chain-wound clock 175
  key-wound clock 174–175
  mounting bolts 176
  seating board 174
steel slot-headed screws 52

taps and dies 31–32
time 6–7
tools
  bench-top anvil 28–29
  bench vice 32
  centre drill 21
  depthing tool 31
  finishing papers and films 34
  five-sided broach 30–31
  four pin vices and one hand vice, selection of 29
  graver 26
  hammer 28
  hand files 25–26
  hand puffer 34
  horological wheel cutters 23
  jobbing drills 21
  lathe
    chuck 21, 22, 23
    collets and chucks 23
    Pultra 1770 lathe 22
    Schaublin 102 lathe 22
    Schaublin vm102 lathe 22
  machine vice 32
  measuring and marking tools 27–28
  oil stone 21
  piercing saw 23–25
  pillar drill 20
  pivot file/burnisher 29–30
  screwdrivers 27
  spotting drills 21
  taps and dies 31–32
transit instrument 7

verge escapement 10

weight, the 168
wheel crossing 73–79
wheels
  Chronos wheel cutting tool 60
  in horological work 59
  making 62–79
  a mandrel 60–61
  marking out 61–62
  Schaublin 102 instrument maker's lathe 60
  terms and dimensions 18
wheels and pinions
  into plates
    end shake 123–124
    five-sided broach 119
    mounting centre and third wheels 120–122
    oil sink cut into the pivot holes 122–123
    pivot hole, drilling 118–119

First published in 2025 by
The Crowood Press Ltd
Ramsbury, Marlborough
Wiltshire SN8 2HR

enquiries@crowood.com
www.crowood.com

British Library Cataloguing-in-Publication Data
A catalogue record for this book is available from the British Library.

For product safety-related questions, contact productsafety@crowood.com.

ISBN 978 0 7198 4386 0

Typeset by Simon and Sons
Cover design by Su Richards
Printed and bound in India by Thomson Press India Ltd

Acknowledgements
So many folk have helped with this publication and my thanks go to them all. Of special mention are: John and Rosalind Middleton, Joel and Tom Simmonds, Myles Crowder, Graham Bemment, and my wife, Gill. Thank you!